THE NEW COSMOS

(*The genie of the pyramid*)

Wes Jamroz

Troubadour Publications

The New Cosmos

(The genie of the pyramid)

Editing: *Dominique Hugon, Patrick Barnard*

Cover illustration: *Sandra Viscuso* (sandraviscuso.com)

Copyright © 2021 by Troubadour Publications. All rights reserved.

No part of this work may be reproduced or transmitted in any form or by any means, electronic or mechanical, including photocopying and recording, or by any information storage or retrieval system without the prior written permission of Troubadour Publications.

Montreal, QC, Canada

TroubadourPubs@aol.com
http://www.troubadourpublications

ISBN: 978-1-928060-15-4

The phenomenal is the bridge to the real

Table of Contents

The Cosmic Matrix 9
The Structure of Mystical Texts 19
A Rendez-vous with Santa Clause 31
The Genie 37
The Secret of the Pyramid 45
Italian Duels 65
The Second Arabian Nights 81
The Craftsmen of Science 95
A Travelling Scientist 103
The Metaphysicists of Copenhagen 109
Tridents, Imp, Gunpowder, and a Cat 115
Spectrum of Matter 131
Who is the Observer? 139
The Western Lover 151
The Dream 165
A Magic Carpet and Aleph 171
Complex versus Real 189
The Entangled Mind 197

The Cosmic Matrix

The Universe is like a corporeal body placed temporarily within the incorporeal Macrocosm.

The Cosmos is a gradient of consciousness. According to the model described in *A Journey through Cosmic Consciousness*[1], it consists of three main segments, which may be illustrated by a three-level pyramid. Each segment belongs to a different stratum of consciousness.

At the top of the pyramid, there is a pyramidion, which represents the original triplicity, the origin of everything. The triplicity is a symbol of cosmic perfection. As such it does not change; its structure is permanent. It contains within itself the entire design, or the cosmic matrix, of the invisible and visible worlds. The cosmic matrix is projected onto the two lower segments of the pyramid. This projection constitutes the descending part of the creation. The projection is realized through the field of cosmic consciousness.

The middle part, in the shape of a truncated pyramid, represents the Macrocosm – the invisible worlds. The Macrocosm contains the symbolic forms of all elements and creatures of the physical world. It also holds the entire sequence of the downloading of the symbolic forms into their physical manifestations. The structure of the Macrocosm changes. However, these changes are not limited by conventional time and space. The changes occurring in the

---

[1] *A Journey through Cosmic Consciousness*, W. Jamroz (Troubadour Publications, Montreal, 2019).

Macrocosm may be stopped, accelerated, slowed down, or partially reversed.

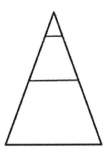

The bottom part, also in the shape of a truncated pyramid, represents the universe, i.e., the physical world. The physical world occupies the lowest zones of consciousness. The physical world is placed within the limitations of space and time.

The process of creation was initiated with the appearance of purely macrocosmic entities during its descending phase. These entities which appeared in the Macrocosm are referred to as *ideas* and *qualities*. At the next stage of the process, the symbols of future earthly creatures and elements appeared. This started with the formation of the symbolic form of the human mind ("Adam"). Afterward, the symbols of animals and plants appeared. The various stages of the descending phase of creation are described in the second chapter of *Genesis*. This phase was completed prior to the appearance of the physical world, i.e., before the Big Bang. This means that, at that time, the physical world was still non-existent.

After the completion of the descending phase, the ascending phase of the process was initiated and included the gradual formation of physical structures. The ascending phase was initiated with the Big Bang. It started with the appearance of space-time. The first physical manifestation of this projection from the Macrocosm was in the form of oscillating fields containing imprints of the future elementary particles.

It is quite interesting that scientists were able to reconstruct the first form of the physical world. They refer to it as physical "nothingness." However, this "nothingness" does not mean that there is nothing. It has been determined that "nothingness" contains bubbling fields within which various shapes of future particles are popping in and out. This bubbling "nothingness" represents the first physical manifestations of the macrocosmic symbols. Scientists coined this bubbling "nothingness" – *quantum vacuum*. The enclosed image illustrates a computer simulation of the quantum vacuum:

Visualization of "nothingness" (Image credit: Derek Leinweber[2])

The sequence of appearance of the physical forms within the universe runs in reverse order to that implemented in the Macrocosm during the descending phase. Namely, after the

---

[2] http://www.physics.adelaide.edu.au/theory/staff/leinweber/VisualQCD/QCD vacuum/index.html.

appearance of space-time, the process followed with the creation of elementary particles, atoms, minerals, plants, and animals. With the appearance of mankind, the deterministic part of the ascending phase described in the first chapter of *Genesis* was completed. Symbolically, this phase of the creation may be envisaged as filling the lower truncated pyramid with physical elements and creatures. At that time, there were no traces of human contribution to the Macrocosm which did not contain yet any "perfected" human minds. With this respect, it may be said that the Macrocosm was still "empty."

The enrichment of the Macrocosm with perfected minds is the ultimate purpose of mankind. This goal may be accomplished by activating certain faculties of the human mind. In an ordinary mind, these faculties remain in their latent states. By activating them, a new or perfected mind is formed. In this way, the human mind is transmuted. A transmuted mind can operate within the invisible zones of the Macrocosm. It enriches the Macrocosm by becoming a part of it. It is in this manner that a New Cosmos is being gradually formed and developed.

The presence of perfected minds within the Macrocosm must keep increasing to overcome the entropy of human existence on the planet. Perfected minds are often referred to as "Real people":

> They are not ordinary men, let alone monks. They know neither rest nor even satisfaction, for they have to make up for the shortcomings of humanity. They are Real people who have experienced being and non-being and have long ago entered a stage of evolution when neither state means anything to them.[3]

---

[3] *The Teachers of Gurdjieff*, Rafael Lefort (Victor Gollancz Ltd., London, 1966, p. 96).

There is a certain minimal critical mass of "perfected minds" that must flow into the Macrocosm. Therefore, each human generation must contribute to that process to sustain the universe. It is this mechanism that keeps the universe alive.

To ensure that the process continues, perfected minds need to populate the various macrocosmic strata. There is a specific pattern of distribution on each level of the Macrocosm that must be completed during specific time periods. These various zones of the Macrocosm may be compared to the set of strings of a musical instrument. The arrival of perfected minds makes them oscillate at "frequencies" specific to their locations within the Macrocosm. Together, these oscillations generate "sounds" which are often referred to as the music of the spheres. It may be said that men may gain immortality by tuning their minds into resonance with this music that cannot be heard.

Are there any practical benefits to be derived from knowing about this process? In other words, is this sort of information of any practical use? Why should anyone bother himself or herself with such seemingly abstract and irrelevant concepts?

Let me add a couple more points to the above description, so that we may start to see more clearly its practical relevance.

As soon as a zone of the Macrocosm is filled in with perfected "earthly" minds, it becomes itself an active element of the New Cosmos. Then —and this is the main point of the entire enterprise— it is projected back onto the ordinary human level. It is in this manner that each new cell of the Macrocosm contributes to upgrading the matrix that serves as a template for the next generations of humans. The evolution of man is realized through the adaptation of the ordinary human mind to this continuously upgraded macrocosmic template.

The changes taking place in the Macrocosm are manifested by the appearance of new concepts and ideas. This is the prime mechanism that initiates new aspirations, new movements, new developments – which are on the unfolding path of human evolution.

It should be possible, therefore, by looking back at our history, to detect traces of this sort of cosmic adjustments. It would be of great interest to identify the traces of such "upgrades." Very often such "upgrades" are associated with a chain of disturbing and stressful events. Being familiar with their origin would allow us to have a much better appreciation of events that, otherwise, may seem to be senseless, irrational, meaningless. In this way, we may gain a much better understanding of what is happening around us. Secondly, recognizing the invisible thread interwoven into these events and happenings would help us diminish, at least partially, their disturbing influence on us.

***

Each new concept is conceived within the Macrocosm. Then, it is projected through the cosmic field onto the zone of consciousness that operates within the physical world. These concepts have a certain degree of complexity. Their essence is such that, at the time of their first appearance, the ordinary human mind can't perceive them. Their understanding depends on the perceptive capacity of the human mind. It takes centuries, or even millennia, before these concepts can be effectively understood and implemented. This means that their discoveries belong to specific periods of the history of humankind. First, they are expressed in their symbolic forms and are recorded in scriptures, tales, or poetry. At the next step of their unfolding sequence, they may be translated into mathematical formulas or expressed as laws of physics. It is at this point, that a new concept is celebrated as a great discovery of the human mind. Afterward, it is permanently instilled in the minds of most people.

Let us take as an example the famous principle of uncertainty. It was formulated by Heisenberg in 1927. The principle asserts that the position and the velocity of an object cannot be exactly measured at the same time. It means that nature imposes a limit below which certain values and measures cease to exist. It is not a matter of not having access to precise enough instrumentation. It is the property of the ... physical world; beyond this limit, an object becomes unmeasurable. When the uncertainty principle was formulated, it sent shock waves among the scientific community. It was one of the first warnings that deterministic science entered onto its diminishing slope. Yet, this seemingly simple and intuitively plausible principle entered the human mind much earlier. It was formulated in the 5th century BC in Greece by Zeno of Elea. It is known as Zeno's paradox. The paradox refers to Atalanta, a famous huntress, who was renowned for her speed:

> To go from her starting point to her final destination, Atalanta must first travel half of the total distance. To travel the remaining distance, she must first travel half of what is left. No matter how small a distance is still left, she must travel half of it, and then half of what is remaining, and so on, ad infinitum. Because of the infinite number of steps required, she can never complete her journey.

Although quite correct logically, Zeno's paradox leads to an obviously ridiculous conclusion. It indicates that there is a limit beyond which such linear thinking is not applicable. Even though it is expressed differently, Zeno's paradox was the first intuitive expression of the Heisenberg principle of uncertainty; it was its equivalent. Over the centuries, philosophers, mathematicians, and physicists could not come up with a satisfactory explanation of this paradox. The difficulty of its understanding was not related to a lack of advanced measuring devices, powerful computers, or sophisticated mathematical tools. Its understanding was being blocked by a certain conceptual barrier. Namely, the human mind could not accept that there might be a gap on the path from the finite to the infinite or from the visible to the invisible. It took nearly twenty-five hundred years to overcome this particular conceptual barrier, at least partially. It was only after the birth of quantum mechanics that scientists were forced to accept such an option. Without quantum mechanics, it would still be impossible to arrive at a satisfactory explanation of Zeno's paradox. It is important to underline the fact that the solution was not the result of a sudden and brief "Eureka" kind of experience. This was a long, gradual, and sometimes tedious struggle with many mental obstacles that stood in the way.

**

The purpose of human life is to contribute actively to the enrichment of the New Cosmos. Only by constructively contributing to the advancement of the cosmic matrix may humans sustain the evolutionary process and, in this way, preserve their existence.

This statement implies that there are continuous changes within the macrocosmic structure. Is there any way to verify this? Is it possible to find out what type of changes humans make to the invisible worlds?

Yes, there are indications of this process. The purpose of this book is to identify them and describe how these changes have affected the human mind over the last several thousands of years.

The Structure of Mystical Texts

> It now appears that research under way offers the possibility of establishing the existence of an agency having the properties and characteristics ascribed to the religious concept of God.
> (*Evan H. Walker*)

It is of great importance for the preservation of the human race that men should be able to grasp the overall concept of the cosmic structure – so that they can correctly discharge their cosmic function. The challenge was and still is, how to make humans aware of their role within the cosmic design. It was for this purpose that the overall scheme and mechanism of the Cosmos have been communicated to men. Of course, to warrant the durability of the message, it had to be described in a highly symbolic way. Otherwise, if limited to a specific language and locally used terminology, its originally intended meaning would quickly be corrupted, misinterpreted, and misused.

Some pieces of information about the macrocosmic structure were disclosed in *Genesis*. *Genesis* contains a basic introduction to the overall process. The composition and the arrangement of some of the stories of *Genesis* describe the macrocosmic design and the process of creation.

The birth of the New Cosmos was marked by the "return of Adam." "Adam" was the first perfected mind or "earthly soul" who returned to the Macrocosm. He was the first element of the New Cosmos. Since that time, the process of developing the New Cosmos has continued. As mentioned earlier, the process consists of the activation of certain faculties of the human mind. The activation of these latent faculties is realized through experiences consisting of a series of ascents into the invisible worlds. This sort of experience is symbolically described in ancient texts and mystical poetry. These texts may be recognized because their inner design is based on a pyramidal structure.

The pyramidal structure was used as a template for many biblical stories. The story of Noah in *Genesis* may serve as an example of such an illustration. Noah's story describes an experience that consists of an ascent that is followed by a descent. Here are the main events experienced by Noah (*Genesis*, 6.10 – 9.19):

1. Noah was a righteous man, and perfect in his generation. Noah walked with God. Noah begat three sons, Shem, Ham, and Japheth.

2. God said unto Noah, ... but I will establish my covenant with thee.

3. Noah went in, and his sons and his wife, and his sons' wives with him, into the ark ... They and every beast after its kind.

4. The waters prevailed and increased greatly upon the earth; and the ark went upon the face of the waters.

5. God remembered Noah, and every living thing, and all the cattle that were with him in the ark; and God made a wind to pass over the earth, and the waters assuaged.

6. Noah went forth, and his sons, and his wife, and his sons' wives with him. Every beast, every creeping thing, and every fowl, whatsoever moveth upon the earth after their families, went forth out of the ark.

7. And God spake unto Noah ... I will establish my covenant with you and with your seed after you.

8. And the sons of Noah that went forth of the ark were Shem, Ham, and Japheth. These three were the sons of Noah and of these was the whole earth overspread.

The first four events describe the ascent; the remaining events are the description of the descent. The story's main theme is the preservation of Noah's family. This is why the story starts and finishes by emphasizing the fact that Noah had sons. In other words, the story is an allegorical illustration of the kinds of experiences needed to sustain humanity.

The ascent is initiated at the base of the truncated pyramid. It is marked as event number one (1).

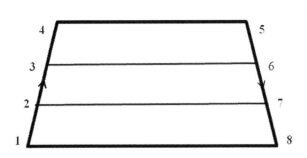

In accordance with the symmetry of the pyramid, the events number one (1) and eight (8), two (2) and seven (7), three (3) and six (6), four (4) and five (5) – are conjugated. The events (2) and (7), for example, refer to God's covenant with Noah. They emphasize the fact that the entire process was implemented following the cosmic design ("my covenant with you").

The next events in Noah's experience define the select sample of creation which was exposed to the macrocosmic matrix projected onto the earth at that time. These are events number three (3) and number six (6). They indicate that the projected matrix was applied to Noah, his family, and select species of animals. They were to serve as seeds for future stages of the evolutionary process.

The next events, i.e., the events number four (4) and number five (5), marked the beginning and the end of the flood, respectively. In mystical texts, an ocean or a sea is often used to indicate higher zones of consciousness. In other words, events number four (4) and number five (5) determined the duration of Noah's exposure to higher zones of consciousness. It was during that time that Noah and his family were exposed to the higher zones of the field of consciousness. In this way, they were "enriched" or "upgraded" – following the renewed pattern of the evolutionary matrix. At the same time, a cleansing of the earthly environment took place. The cleansing is allegorically indicated by the changes that occurred in the worldly environment that remained outside of Noah's ark. The earthly environment was cleansed from elements that would interfere with the new evolutionary pattern. Outside the ark, the situation corresponded to a "dark age." Afterward, the process entered its descending phase. The descent consisted of three events that were parallel to the ascending ones. Upon Noah's return, the entire environment was changed – allowing for the constructive integration of a renewed evolutionary pattern.

The last event, i.e., event number eight (8), emphasizes the fact that the entire experience brought humanity, allegorically referred to as Noah and his family, onto a higher level of being. The seed of a new "civilization" was planted.

This pyramidal representation underlines the main feature of developmental texts. Although the initial and the final steps are located at the base of the pyramid, the state of the "traveler" at the conclusion of the "journey" is different than the initial one. This is shown by having the point of descent (the event #8) in a different place than the position of the initial point (the event #1). As a result of this experience, the mind of the traveler has been changed. The outcome of such an experience always leads to a gain in knowledge and understanding.

When Noah's story is read sequentially, it may seem to be repetitious, incoherent, and a bit messy. Once the inner structure of the text is seen, then it is easier to recognize its intended meaning.

Many mystical texts are structured in this kind of pyramidal schema. It is a very convenient representation because it allows describing the relationship between several levels of consciousness. Starting with *Genesis*, this literary form was adopted in mystical literature worldwide.

Let us take another example. Here is a poem written by Fakhruddin Iraqi, a 13th century Persian poet:

When the sea breathes they call it mist;
When mist piles up they call it clouds.
It falls again, they name it rain;
It gathers itself and rejoins the sea.
And it is now the same sea it ever was.[4]

---

[4] *Divine Flashes*, Fakhruddin Iraqi; translated by William C. Chittick and Peter Lamborn Wilson (Paulist Press, Mahwah, NJ, 1982, p. 78).

This poem describes a process in which drops of water experience several transformations. In this poetical illustration, the sea represents the top layer of the Macrocosm; the clouds are a symbol of the physical world. In this case, the evolutionary process is described from the macrocosmic point of view; it is a reverse view of that illustrated in Noah's story. Therefore, the first part describes the descent. The first verse marks the beginning of the process: the water of the sea is transformed into "mist." In the second verse, the mist is transformed into "clouds." This is followed by a return when the clouds are transformed into "rain." In the fourth verse, the rain rejoins the sea. In other words, the poem illustrates the process of enrichment of the Macrocosm or the formation of the New Cosmos. In a graphical illustration, this poem could be presented as a truncated pyramid.

The last verse of the poem states that the sea is now the same as "it ever was." It was these sorts of expressions that confused many interpreters of mystical poetry. It was somehow assumed such descriptions indicated a ... circular cycle of life, a fatal cycle of birth and death without progress. Instead of a pyramidal structure, these types of arrangements have often been presented as a circular or a ring composition. The circular composition indicates that the final state is identical to the initial one.

The last line of Iraqi's verse is there to trigger a reaction from the reader. Namely, if at the end of a series of quite sophisticated transformations, there is no difference between the initial and the final states of the sea – so what is the purpose of these experiences? If this kind of "journey" is a symbolic illustration of human life, then what is the purpose of life? The purpose of this particular verse is to prompt such questions and force the reader to ponder on them. In this way, the poem helps to identify the gamut of assumptions that are ingrained in the reader's mind. This mechanism is nicely explained by Kingsley Dennis, a contemporary writer:

There have been a significant number of developmental texts that were constructed in a deliberate manner to oppose a rational and sequential approach. Their purpose is to affect a change in the reader's perception in order to trigger an altered comprehension of reality. Such works aim for a comprehensive and integral understanding.[5]

Although the form of the "sea" is the same, its inner state has been changed. Namely, the "sea" has been enriched by its experience of being a cloud. The enrichment may seem insignificant, nevertheless, it quite precisely reflects the function of humanity and the purpose of human existence. Saadi of Shiraz, a Persian poet and a contemporary of Iraqi, provided a further hint that may help to grasp the meaning of such enrichment. He used a poetic convention according to which a pearl is a transformed raindrop:

> A drop which fell from a rain-cloud
> Was disturbed by the extent of the sea:
> 'Who am I in the ocean's vastness?
> If IT is, then indeed I am not!'
> While it saw itself with the eye of contempt
> A shell nurtured it in its bosom.
> The heavens so fostered things
> That it became a celebrated, a royal Pearl.[6]

---

[5] *The Modern Seeker*, Kingsley Dennis (Beautiful Traitor Books, 2020, p. 113).
[6] Translation by Idries Shah, *Learning How to Learn* (Harper & Row Publishers, San Francisco, 1981, p. 136).

We may recognize that the "returning" drops of water are an allegorical illustration of the "enriched" earthly minds that contribute to the development of the New Cosmos. They are like "royal pearls" in the ocean's vastness. This kind of change is graphically incorporated within the shape of the pyramid. The shape of the pyramid indicates that although the initial and the final states are placed at the same level within the pyramid, they are not overlapped. They are different.

***

The biblical records constitute a literary account of the initial structure of the Macrocosm. Although these records were composed some thirty-two centuries ago, their content applies to the very beginning of humanity. According to the biblical account, in the beginning, there were no earthly souls in the Macrocosm. There were no traces yet of a New Cosmos.

The next insight into the macrocosmic structure was provided by Mohammed's experiences during the Night Journey. This was the first account that confirmed a human "contribution" to the structure of the invisible worlds. The contribution has been allegorically described as the presence of "perfected human minds" within the macrocosmic structure. This adjustment of the Macrocosm was made between the time of "Adam's return" and the time of Mohammed's journey (621 AD).

Here is an abbreviated account of Mohammed's experiences:

During the journey, Mohammed was guided by the archangel Gabriel. First, they reached the lowest level of the Macrocosm. The lowest level of the Macrocosm corresponds to a level of consciousness that is associated with Adam, the symbolic father of humanity. This layer of consciousness is often referred to as the first heaven.

Then they ascended to the second heaven. The second level is associated with Jesus and John the Baptist, the sons of two sisters. This layer of consciousness is symbolically referred to as the second heaven.

The third layer recorded in Mohammed's journey is associated with Joseph, son of Jacob.

The fourth layer is associated with Idris, an ancient prophet and patriarch. He is sometimes identified with the biblical Enoch and with Hermes Trismegistus.

The fifth layer is marked by the presence of Aaron, the elder brother of Moses.

The sixth layer of the Macrocosm is that of Moses.

The seventh layer, or the seventh heaven, is presided over by Abraham.

In summary, the structure of the New Cosmos at the time of Mohammed was as follow:

7th heaven (Abraham)
6th heaven (Moses)
5th heaven (Aaron)
4th heaven (Idris)
3rd heaven (Joseph)
2nd heaven (Jesus and John the Baptist)
1st heaven (Adam)

Mohammed's account was the first indication of the formation of the New Cosmos. At that time, the structure was divided into seven layers. Of course, this was a symbolic illustration that corresponded to the visible configuration of "heavenly" planets. The fact that the seven heavens coincided with the seven planets boosted up interest in medieval astrology. As each layer was associated with one of the biblical characters, it allowed for drawing astrological links between them and the planets. As we will see later, this correlation provides quite an interesting insight into the overall concept of astrology.

The biblical characters associated with those various macrocosmic levels are referred to as prophets, messengers, or teachers. They represent the "perfected" human minds that formed the foundation of the New Cosmos. Their arrangement within the macrocosmic structure indicates a certain hierarchy among them. However, except for the first one (Adam), this hierarchy does not reflect the chronological order in which these individuals appeared on the earth. Otherwise, their arrangement would be Adam, Idries, Abraham, Joseph, Moses, Aaron, John the Baptist, and Jesus. It seems that some other factors determined the hierarchical arrangement.

Mohammed's account provides another interesting piece of information about the structure of the New Cosmos. Namely, in that hierarchical structure there is an upper limit for the original "heavenly" minds – symbolically referred to as *qualities* and *attributes*. Their limit is set at the symbolic layer designated as the 7th heaven. This is why, Gabriel –who represents a certain original *quality*– could not pass beyond that limit. He had to stay there, while Mohammed could travel beyond. This indicates that a fully developed human mind may pass beyond that limit and experience the 8th heaven. The 8th heaven is represented by the pyramidion that is placed on the top of the truncated pyramid. Within that zone, the human mind is dissolved. It is absorbed within the fabric of the Absolute.

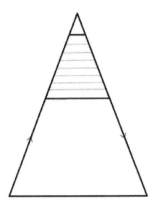

The outline of the New Cosmos was divided into seven zones. The names associated with the different zones are the symbolic representations of the newly activated layers of the human mind. Before their activation, these layers remained in their latent states. These levels are symbolically referred to as "heavens." In the language of astrology, this means that the "heavenly" constellations are not fixed; their qualitative content keeps changing in accordance with the development of the human mind. This means that by developing their minds, humans can change the quality of the "heavenly" constellations and, therefore, can affect their destiny. In other words, human destiny is not determined by fixed planetary arrangements. Destiny is determined by the quality of human efforts. This observation provides an interesting insight into the overall concept of astrology.

The evolution of the human mind changes the macrocosmic structure. This structure is then projected back onto the physical world. The changes within the Macrocosm are to be absorbed by the earthly minds. The mystical texts are designed in such a way as to prepare the human mind for such evolutionary adjustments. These adjustments have been communicated in their symbolic forms through the introduction of new concepts and ideas.

Scriptures, mystical poetry, and certain fairy tales have served as a conduit for these new concepts and ideas.

A Rendez-vous with Santa Clause

> Those dupes of intellect and logic die
> In arguments on being and not being.
> Go, ignoramus, choose your vintage well –
> From dust like theirs grow none but unripe grapes.
> (*Omar Khayaam*)

"So, your interest is in sciences and their role in human development?"[7]

It was early December in 1991. I was sitting with Santa Clause in a small café on Monkland Avenue. I had met him on the street, a few moments before, in that freezing and snowy evening. He was carrying an incredibly huge bag and he looked very tired. I had asked him whether I could help him. He had said that he was fine, but if I could invite him for a cup of coffee, he would appreciate it very much.

Somehow, I had started to talk about my interests in the latest scientific discoveries and the possibility that, soon, science would be able to explain the basic questions related to the origin, purpose, and destiny of man.

"You should not," he said, "mix up these trials and errors of the human intellect with real knowledge. These, as you call them rational sciences and with which you are so preoccupied, are very much

---
[7] This conversation was originally published by *The London Magazine – An-Naar*, Issue N° 1, 1991. (Published irregularly and infrequently, *The London Magazine* was distributed among a limited number of subscribers.)

limited. They have little to do with human development. As a matter of fact, they are themselves limited by the shortcomings of human nature. Arrogance, pride, greed –all of these– form sort of a veil, which separates humans and their intellectual activities from the true reality."

"But I am not talking about ... psychology or anthropology or this type of staff" – I interrupted him, "I mean 'hard' sciences such as modern physics, astronomy, and cosmology."

"That is exactly what I am referring to" – came his reply. "You see, you yourself cannot even imagine that the human mind itself can affect the results of materialistic investigations."

I had to admit it was true. Such a possibility had never occurred to me.

"Let me give you some examples," Santa continued, "as you know, modern physics precisely defined the limits within which all its exercises and speculations have to be enclosed. The famous principle of uncertainty defines the limits of what can be observed and measured. Therefore, there are things that by definition cannot be 'seen' by scientists and their instruments. The second law of thermodynamics says that disorder will tend to increase if things are left by themselves. This means that there is no chance for any materialistic evolution without some intervention from outside of the system. And finally, the velocity of light is the highest possible speed for the transmission of any kind of energy. What 'incomplete science' -as we should call it- does not know and is too arrogant to even consider is that some of these mechanical laws were known centuries ago. For example, ... may I have one more cup of coffee, please?"

A waiter fulfilled his request immediately. I was impatiently watching Santa as he was slowly sipping his coffee.

"Do you know that the limiting velocity of light was suggested in the 11th century by Avicenna?"

"No, I have never heard of that."

I knew that the velocity of light was measured for the first time by the Danish astronomer Ole Roemer in 1676. His measurements were based on the observations of Jupiter's moons. But it was not until 1905, that Albert Einstein concluded that the velocity of light was the highest possible speed in the universe.

"It does not take a genius to realize" – Santa looked at me with a smile, "that we are all using a sort of energy which is transmitted faster than ... light."

I could not mask my excitement. Thoughts were speeding up in my head ...

"Your thoughts ... " – he said slowly, "your thoughts can be transmitted faster than light. What is the distance from the Earth to the Sun?" he asked.

"Approximately ... eight light-minutes."

"And to the Alpha Centauri?"

"Something like ... four-and-half light-years."

"So?" – asked Santa, "how much time would it take for your thoughts to travel over these distances? And what would happen if you tried to illustrate these two trips on space-time diagrams? Wouldn't it be possible to imagine, that the absolute past and the absolute future of these two events could become ... everlasting present?"

I was a bit confused. Thoughts are not ... material, I thought, nothing and no-one can detect them, never mind ... measure them.

"Those who cannot measure them" – Santa obviously detected my thoughts, "simply ... cannot measure them. Scientists are too proud of their own 'achievements' to even consider that option."

I could not imagine that!

"You see, our thinking process is much more complex than you can imagine. Of course, there are some limits. There are limits because of your incompleteness. And the limiting factor is ... your imagination. Imagination can be developed, qualitatively and quantitatively. And this is in direct relation to ... knowledge. The true complete knowledge. Now, the problem with this incomplete science is that as long as it sticks to its limits, there is no hope for any real progress."

"But what about such discoveries as ... nuclear energy, electronics, lasers? These are the real facts, practical results, aren't they?" – I was trying to defend my point.

"Your lack of perception amazes me" – came his answer, "I thought that you were interested in human development, in knowledge. But, it seems, that you are just fascinated with materialistic gains. Of course, technological inventions are valid, even necessary if used and applied in the right context. However, they are only ... by-products, results of mistakes, coincidences, and misinterpretations. For example, you yourself know very well that the laser theory was written down by Einstein – but he himself did not realize its practical implications. Until his last days, he was working on the unified theory, the task which he failed to achieve. The history of scientific discoveries is full of examples like this. Scientific discoveries are used as a smokescreen to hide the fact that ... May I smoke here? I have been warned that in this country I may be prosecuted for smoking."

"It's o.k. We are in the smoking section," I assured him.

Santa lit his cigarette. Then he continued:

"For centuries, no real progress has been made. As you know, in this decade scientists are going to spend millions in the race to discover the fundamental blocks of matter. What do you think, are they going to succeed?"

"Oh, yes. If they can find the top quark,[8] the only missing element in the Standard Model, then everything ... "

"I would expect" – Santa interrupted me, "that you are capable of learning something from the history of scientific discoveries. You do not need a crystal ball to predict that this race of relentless ambition and monumental egos will never end. What is going to happen is that a whole new layer of matter will be eventually discovered. This will force a new crisis in science, followed by a far more complicated set of new theoretical models, particles, etc. You may expect the arrival of these future particles in the form of preonic strings or something like that. But as long as scientists can convince their power-greedy sponsors, that one day they will be able to create for them ... new universes (!) – this race will go on. At the same time, people do not realize that since the beginning of time there has existed complete, precise, and objective knowledge. All the parameters, all the possibilities, and elements of the cosmic matrix ... are known. There is no need for any further speculations. It is only a matter of its proper use and correct application.

Now I must return to my work. I leave you a present."

He took out from his bag a nicely wrapped parcel, handed it to me and said:

---

[8] The top quark was discovered in 1995, i.e., four years after this conversation took place.

"You will find in it a book called *One Thousand and One Nights*. Read it, but do not try to interpret it – right now you cannot yet. You will not find answers to your questions in it. However, you may learn how to improve your imagination. Farewell!"

And he was gone.

"Would you like more coffee?" – the waiter was smiling to me.

"Coffee? No ... double vodka, and no ice, please."

The Genie

The strength of the genie comes from being in a bottle.
(*Richard Wilbur*)

One Thousand and One Nights, or the *Arabian Nights* as it is commonly known in English, is a collection of stories of uncertain date and authorship. The origin of most of the tales, however, may be traced to Indian, Persian, and Arabic cultures. The stories collected in the *Arabian Nights* illustrate a unique world of extraordinary riches, strange places, magical creatures, and improbable happenings.

The first known reference to the *Arabian Nights* is a lost Persian book of fairy tales called *A Thousand Legends*. This book was translated into Arabic about 850 AD. The first reference to the Arabic version under its full title *One Thousand and One Nights* appeared in Cairo in the 12th century. Yet, many stories included in the *Arabian Nights* were well known for centuries in their oral versions. For example, the tale of Aladdin and his wonderful lamp has been found in oriental manuscripts in South India, Burma, and Albania. It also exists in oral versions in Greek, Czech, and Mongolian.[9]

One Thousand and One Nights was first published in the Western world by Antoine Galland, a French orientalist. The twelve volumes of Galland's *Mille et Une Nuits* were printed between 1704 and 1717. The first complete translation in English was published by John

---

[9] *World Tales*, Idries Shah (The Octagon Press, London, 1991, p. 268).

Payne in 1882-1884. This was a nine volumes series printed in a limited edition of five hundred copies. The most celebrated English translation was made by Sir Richard Burton. It was published in sixteen volumes in 1886-8.[10]

One Thousand and One Nights is considered the Mother of Records, Prototype of Tales, or Source of Tales. These terms are derived from the *abjad* system.[11] In other words, the overall design of the *Arabian Nights* may be looked at as a template that contains the overall pattern of the development of the human mind. Like the biblical stories, the pattern of each of the stories is based on a pyramidal structure.

All tales in the *Arabian Nights* are embedded within a single frame-story. Consequently, the first and the last episode of the frame-story form the boundary within which all other tales are enclosed. In accordance with the pyramidal structure, the last episode takes place within the circumstances described in the first story:

> The frame-story presents King Shahriyar who, upon discovering his wife's unfaithfulness, kills her and those with whom she has betrayed him. Afterward, he decides to marry and kill a new wife each day until no more candidates can be found.
> The king's vizier has a daughter, Shahrazad. Shahrazad devises a scheme to save herself and the others. She convinces her father to marry her to the king. After the marriage ceremony, she tells her husband a story, leaving it incomplete and promising to finish it the following night. The king, eager to hear the end, delays her execution for another day. On the next night, she continues her story by introducing a new story

---

[10] From "Introduction" to *Tales from the Thousand and One Nights* translated by N.J. Dawood (Penguin Books, London, 1973).
[11] *The Sufis*, Idries Shah (The Octagon Press, London, 1989, p. 175).

within the previous story – without finishing it. She keeps telling her stories night after night. Finally, the king is so impressed with her intelligence, eloquence, and the lessons embedded in her stories that he decides to abandon his cruel plan.

The uniqueness of the *Arabian Nights* is that its overall design reflects the cosmic structure recorded in Mohammed's account. Namely, the frame-story is like a structure that encompasses the Macrocosm and the physical world. The multiplicity of the macrocosmic zones experienced by Mohammed is reproduced in the *Arabian Nights* by multiple travels of the same character. For example, Sindbad goes on seven consecutive voyages. Then there is "The Tale of the First Dervish" followed by "The Tale of the Second Dervish" and "The Tale of the Third Dervish." Similarly, "The Tale of the First Girl" is followed by "The Tale of the Second Girl," and so on.

In these stories, the main character goes on a long journey during which he encounters unusual people, strange creatures, and magical events. These events symbolically indicate encounters with the invisible worlds. As a result of these experiences, the hero is greatly "enriched." Afterward, he returns to his home where he shares his riches with friends and fellow citizens and tells them about his adventures. Then, he goes for another trip that follows the same pattern. All these multiple travels are linked together.

Graphically, the structure of the *Arabian Nights* may be presented as multiple pyramids enclosed within the main pyramidal frame. The main pyramid represents the frame-story; while all other tales are like smaller pyramids that are inserted inside that frame:

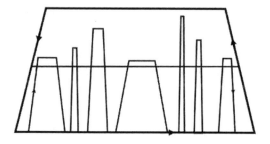

The lower part of the frame-pyramid represents the physical world; the upper part illustrates the Macrocosm. The "magical" events are marked by the smaller pyramids encroaching into the upper part of the frame-pyramid.

There is another factor that links the *Arabian Nights* to the invisible worlds. It is the appearance of a genie (also known as *jinni* or *jinnee*). The appearance of a genie is an important feature of the *Arabian Nights*. Although the first appearance of a genie may be traced to the Talmudic story of Solomon and Ashmedai and the corresponding accounts in the *Koran*, it was the *Arabian Nights* that immortalized this extraordinary creature in world literature.

The appearance of a genie marks the extraordinary interventions which interfere with the hero's travels. At these moments, the course of the journey changes drastically; it diverges from its expected linear pattern. Once changed, the journey continues. Then, another intervention interferes and once more confuses the originally intended plan.

The interventions are a symbolic representation of events that originate within the invisible worlds. Genies' interventions are the moments when the hero's mind experiences an impulse from the macrocosmic matrix. It is a genie that brings in a macrocosmic element. It is in this way that genies add a mysterious dimension to the tales.

A genie is a creature of great interest to us. As we will see later, genies served an important role in the development of the human mind; they greatly contributed to the enrichment of human imagination. It was a genie and his companions that implanted in the human mind the concept of a mysterious device allowing to bridge the ordinary physical world with the invisible worlds.

In the *Arabian Nights*, there are many episodes involving genies. For example, here is a description of an encounter with a genie taken from the tale entitled "The Fisherman and the Jinnee":

> The fisherman removed the lead with his knife and again shook the bottle, but scarcely had he done so when there burst from it a great column of smoke which spread along the shore and rose so high that it almost touched the heavens. Taking shape, the smoke resolved itself into a jinnee of such prodigious stature that his head reached the clouds, while his feet were planted on the sand. His head was as huge as a dome and his mouth as wide as a cavern, with teeth ragged like broken rocks. His legs towered like the masts of a ship, his nostrils were two inverted bowls, and his eyes, blazing like torches, made his aspect fierce and menacing.[12]

As far as our tale is concerned, genies have some features which are of interest to us. Namely, genies may appear in different forms

---

[12] *Tales from the Thousand and One Nights*, N.J. Dawood (Penguin Books, London, 1973, p. 80).

and shapes. For example, in the tale entitled "Khalifah the Fisherman," genies appear in the shapes of grotesque apes:

> Khalifah is a poor fisherman. One day he catches a one-eyed, lame ape. He is about to beat it when the ape tells him to throw his net again instead. This time Khalifah brings in a larger ape of even more grotesque appearance. The ape's eyelids are darkened with kohl, his hands dyed with henna, and he wears a tattered vest about his waist. The ape tells him to cast his net again. After casting his net a third time, he captures another ape. The third ape's hair is red, his eyelids lengthened with kohl, his hands and feet dyed with henna, and he wears a blue vest around his waist. This ape also tells him to cast his net again, and this time he catches fish. The ape tells Khalifah that he must take the fish to Master Abu al-Sa'adat and offer it to him. It works, and Khalifah becomes a wealthy man.[13]

Then, there are various grades of genies. Their ability to perform tricks corresponds to their respective grades. The existence of various grades of these magical creatures is indicated in the tale "The Tale of Ma'aruf the Cobbler." For example, here is a genie explaining the hierarchal structure of the invisible world which, we may notice, is in a pyramid-like shape. *Imps*, for example, occupy the lowest level; *giants* are a bit higher within that hierarchical structure. A *lord* is at the top:

> Amazed at the apparition, Ma'aruf asked, Who are you?
> "I am Abdul-Sa'adah, the slave of the ring," replied the jinnee.

---

[13] *Ibid*, p. 303.

"Faithfully I serve my master, and my master is he who rubs the ring. Nothing is beyond my power; for I am lord over seventy-two tribes of jinns, each two-and-seventy thousand strong: each jinnee rules over a thousand giants, each giant over a thousand goblins, each goblin over a thousand demons, and each demon over a thousand imps. All these owe me absolute allegiance, and yet for all my power. I cannot choose but to obey my master."[14]

We can also find a reference to such a hierarchy among genies in the tale "Aladdin and the Enchanted Lamp." For example, the power of the "slave of the lamp" is much superior to that of "the slave of the ring":

Aladdin rubbed the ring and a genie appeared, saying: "I am here. Your slave stands before you."
Aladdin rejoiced at the sight of the genie, and cried: "Slave of the ring, bring me back my wife and my palace with all that it contains."
"Master," the genie replied, "that which you have asked is beyond my power, for it concerns only the slave of the lamp."[15]

But let us leave the genies here. We will meet these creatures on several occasions later on. There are several other events we must attend to – before our next encounter with one of them.

---

[14] *Ibid*, p. 393.
[15] *Ibid*, p. 223.

The Secret of the Pyramid

> For every trick of imagination there is a reality of which it is a counterfeit.
> (*Khaja Ahrar*)

No structures have raised man's curiosity as much as the Egyptian pyramids. Hundreds of books have been published about them. There is no doubt that hundreds more will be written in the future. Their shapes, locations, and designs arouse the imagination of architects, archeologists, mathematicians, occultists, and others. There are plenty of tales told about the pyramids' mysteries which are related to the various striking intricacies of their geometrical design. Their geometry involves the use of divine proportion or the golden ratio, the Fibonacci series of numbers as well as many features which are widely believed to have deeper symbolic meanings.

The Great Pyramid at Giza is probably one of the most awe-inspiring structures ever built by man. The pyramid's structure has many puzzling details. One of them is related to its pyramidion. Although from a distance the pyramid looks like a perfect four-sided triangular structure, in reality, it is a truncated pyramid. The top of the Great Pyramid is a small flat area, no longer than 30 feet. We may notice that the shape of the Great Pyramid has the same form as that used to represent the structure of the Macrocosm:

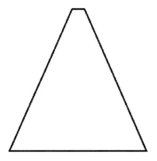

It is believed that the original pyramid had a capstone in the shape of a pyramidion. However, the pyramidion is not there.

There have been many speculations and guesses as to what happened to the capstone. Or if there even was such a capstone. Sometimes the ancients did not complete their temples or monuments to symbolize the imperfection of the mundane world. Some believe that the capstone was made of gold or that it was covered by a layer of gold. It has also been suggested that the capstone was stolen.

The shape of the pyramid inspired many designers, architects, and even politicians. For example, an image of the truncated pyramid was incorporated into the design of the Great Seal of the United States. It appears on the reverse of the Seal and its capstone, in the shape of a triangle, is suspended above the pyramid:

The Great Seal was first used in 1782. However, a die for the reverse of the Seal was not made at that time. There was some hesitation among the members of the US administration to use it because of its supposed controversial symbolism. As a result, the reverse of the Seal remained practically unknown until 1935. In that year, President Franklin Roosevelt decided to place both sides of the Great Seal on the one-dollar ($1) bill. The appearance of the truncated pyramid on the dollar bill led to a renewed interest in its symbolism. Since then, the elements of the design have attracted the attention of various occult and secret societies.[16]

***

Yet, one of the mysteries of the pyramid has so far escaped the scrutiny of pyramidologists and Egyptologists. It is related to a concept that was hidden within the design of the Great Pyramid of Giza but incorporated in its latent form; it was made invisible. Following the overall cosmic dynamics outlined in the previous chapters of this book, this particular concept could not be comprehended by the ordinary men who lived at that time. At that time, it could not be understood because the rational mind was not ready yet for its correct assimilation. It was like a seed implanted in the overall shape of the pyramid. It could not be seen before its germination. It had to remain hidden till the time when the human consciousness would have evolved to a level at which it could not only be perceived but understood and applied. This concept was needed to bring the human mind onto the next level of its evolutionary growth.

---

[16] *Founding Fathers, Secret Societies*, Robert Hieronimus (Destiny Books, Rochester, Vermont, 2005).

The most interesting thing about this concept is the sequence of events that marked the path leading to its discovery. At the various stages of this sequence, many individuals came close to it, nearly touching it, yet – they were incapable of recognizing it. Each event in the sequence involved specific circumstances that included a certain individual, a place within a particular cultural environment, and a stimulating ambiance. Only when all these factors were aligned in a certain manner, the germination of the concept could be advanced. It took several stages, which were separated by centuries before the secret was discovered.

Let us start with a bit of the history that shaped the background of the discovery of this particular secret. It may be traced back to ancient Egypt. It was recorded, in its latent form, in one of the Egyptian papyri. It was encoded within a seemingly marginal problem of algebra.

Among the Egyptian papyri, there is one scroll that contains the calculation of the volume of ... a truncated square-based pyramid. The history of the discovery of this ancient item is itself as a tale from the *Arabian Nights*. It seems that a genie was involved in it from the very beginning. Here is an excerpt from an article that refers to the circumstances which led to the discovery of that papyrus. The article was published in the tourist magazine *Tour Egypt*, in August 2001:

> It was 1871. Grave robbers Ahmed Abd el Rasul, his brother Mohammed, and their accomplice were walking along a path on the face of a cliff in Deir el Bahri, high above the ruins of Queen Hatshepsuts temple. Ahmed suddenly observed a dark area hidden behind a large boulder. Upon closer inspection, he saw a small opening that was exposed just enough to catch the eye of an experienced tomb robber. The test he performed was simple enough: he tossed a rock into the opening and was

rewarded with a long pause before hearing a far-off thud that confirmed his hopeful suspicions. This was an ancient shaft that might lead them to fantastic riches.

Once the men opened up the surface hole, Ahmed went down into the shaft.

Time passed, but the two thieves heard nothing from their leader below. Suddenly a terrifying scream emerged from the shaft, followed by Ahmed hastily clambering up the rope. Gripped with fear, he told his cohorts of his brush with an afrit, a malevolent demon that villagers believed sometimes dwelled in ancient tombs. The looters left in a hurry. Sure enough, proof of the afrit came the following day when villagers detected a nauseating stench on that area of the path, the telltale sign of an angry afrit whose resting place had been disturbed.

During the course of several years following the incident, some extraordinary artifacts slowly turned up at bazaars, auction houses, and in private collections. This set off a stream of rumors that someone must have discovered a treasure-filled royal tomb. Some of the most remarkable artifacts being sold included ushabtis (small blue statuettes) engraved with the name Pinedjem, a 21st dynasty pharaoh, as well as illuminated papyri in unusually impressive condition.

This, in turn, set off an investigation by the Egyptian Antiquities Service, in the spring of 1881. An investigator disguised himself as a wealthy collector and went to Luxor in an attempt to lure the looter out into the open. Eventually, Mustapha Aga Ayat, a Turkish dealer, offered to sell the investigator a royal ushabti, which obviously came from a looted tomb. But justice did not yet prevail, as Ayat was a consular agent for Belgium, Russia, and Britain, giving him diplomatic immunity. However, enough information was obtained to lead the investigator back to Ahmed Abd el Rasul, and the trio of thieves was arrested, questioned, and tortured. But even the severe beatings did not shake them from their agreed-upon story that on the night in question they were merely looking for their lost goat. No one knew about the ropes and digging apparatus they had been carting along with them.

However, the torture managed to create discord among the thieves, who argued about who was tortured the most and was therefore deserving of the greater portion of the treasure. Since most of their neighboring villagers' families had made a living from robbing tombs for centuries, as had the Abd el Rasul family, Mohammed feared that someone, including one of his partners, might turn them in, and he could end up taking the blame. Shrewdly, he decided that the only way to save himself was to be the one to turn in his own partners, which he did in July of 1881.

Mohammed told the local Qurnan governor that Ahmed had found the royal burial site. He confessed that he and Ahmed had created the foul smell of the afrit by killing a donkey and throwing its carcass into the tomb in order to keep other villagers, as well as their partner, away. Mohammed and Ahmed had been looting valuable artifacts from the tomb since then, occasionally putting a few at a time on the market to keep suspicions down and prices up.

A heavily armed official from the Antiquities Service feared for his life when he traveled to Luxor to survey the scene, knowing full well that he could trust no one in the small village since every villager would have gladly killed him rather than lose such a valuable cache. Even after all of the looting committed by the Abd el Rasul brothers, the tomb still contained incredible riches including more ushabtis, alabaster vessels, papyrus scrolls, trinkets, and much, much more. ...

While Mohammed may have thought that crime paid, he soon found out that being a stool pigeon paid even better. He was given a sizable reward and then hired by the Antiquities Service as a foreman. However, years later, when he showed archaeologists a site containing 150 mummies of high priests belonging to the great temple of Amen-Re at Karnak, he was fired. Authorities believed that he already knew about the tomb and had been looting it for some time. So much for Mohammed's honorable career.[17]

One of the stolen items was a papyrus, which was sold to the Russian Egyptologist Vladimir Golenishchev in 1893, by one of the Abd el Rasuls brothers.

Golenishchev is known for his contribution to the studies of the ancient Egyptian religion. He organized expeditions to Egypt on his own account, made archeological excavations, collected written monuments and pieces of art. He approached the Egyptian monuments from the point of view of the development of religious beliefs. He was interested in the changes in various beliefs, ideas, and rites through the centuries. Since 1886, he was the Curator of the Egyptian and Assyrian collection of St. Petersburg's Hermitage. In 1913 Golenishchev donated his collection, including the above-mentioned papyrus, to the Museum of Fine Arts in Moscow in exchange for an annuity. After the October Revolution in 1917, however, his annuity was canceled. The papyrus remains in the Moscow museum.

The text on the papyrus was most likely written down in the 13th dynasty and was based on older material probably dating to the 12th dynasty of Egypt, i.e., around 1850 BC. Its mystery became known only after the text inscribed on the papyrus was translated in 1930. It was then that the scholarly world learned just how mathematically advanced the ancient Egyptians had been. Since then, the Egyptian

---

[17] "Where Have All The Pharaohs Gone?" Anita Stratos, *Tour Egypt*, Vol. II, August 1st, 2001 (http://www.touregypt.net/egypt-info/magazine-mag08012001-magf4a.htm).

papyrus has been known as the "Moscow mathematical papyrus." It contains twenty-five mathematical problems.

The fourteenth problem recorded in the Moscow papyrus contains an example of how to calculate the volume of a truncated pyramid. The problem applies to a specific set of dimensions, i.e., 6 units for the vertical height of the truncated pyramid, 4 units for the length of the side at the base, and 2 units for the length of the side at the top:

> You are to square the 4; result 16. You are to double 4; result 8. You are to square this 2; result 4. You are to add the 16 and the 8 and the 4; result 28. You are to take 1/3 of 6; result 2. You are to take 28 twice; result 56. See, it is of 56. You will find it right.[18]

In the language of today's algebra, the above text may be "translated" into the following set of symbols:

You are to square the 4; result 16:	$(4^2) = 16$
You are to double 4; result 8:	$(2 \times 4) = 8$
You are to square this 2; result 4:	$(2^2) = 4$
You are to add the 16 and the 8 and the 4; result 28:	$(16 + 8 + 4) = 28$
You are to take 1/3 of 6; result 2:	$(1/3 \times 6) = 2$
You are to take 28 twice; result 56 :	$(2 \times 28) = 56$

---

[18] "Four Geometrical Problems from the Moscow Mathematical Papyrus," Battiscombe Gunn and T. Eric Peet, *Journal of Egyptian Archaeology* (Vol 15, 1929, p. 167-85).

The entire text contains a six-step procedure for the calculation of the volume (V) knowing the vertical height (h), the lengths of the base (a), and the top (b) squares. This text indicates that the Egyptians knew the correct formula for obtaining the volume of a truncated pyramid. The discovery of this knowledge is described as "breath-taking" and "the masterpiece of Egyptian geometry."

Today, this procedure is equivalent to the following equation which is known to every student of elementary geometry:

$$V = 1/3\, h\, (a^2 + ab + b^2)$$

In other words, the six sequential lines of the Egyptian text can be substituted by a half-line long "holistic" concept.

At the time when the papyrus was written, the use of algebraic notation was not known. This is why the Egyptian text did not use mathematical symbols. The algebraic notation was introduced by medieval Arab algebraists at the beginning of the 15th century. This means that it took over three millennia before the text presented in the left column above could be presented in the algebraic notation shown in the above formula. Let us notice that over three thousand years were needed for the human mind to adapt to this rather simple notation. This was not related to the need for any great discovery or great invention or some sophisticated calculations. It was simply a matter of introducing a certain concept that would be acceptable by the human mind.

The Egyptian formula contains within itself a "mysterious" element that was needed for the development of modern physics. In the Egyptian version, however, this mysterious element was still in its latent form and remained hidden. It was there but it was not visible. It contained a secret about an "invisible pyramid" hidden within the design of the Great Pyramid of Giza and was indicated symbolically by the pyramid's truncated shape. The mysterious element inserted in this mathematical problem was like a *genie* from the *Arabian Nights* trapped in a bottle:

> ... Inside had been imprisoned a fearsome genie; and the bottle had been cast into the sea by Solomon himself so that men should be protected from the spirit until such time as there came one who could control it, assigning it to its proper role of service of mankind.[19]

This particular Genie was trapped within a small "bottle" in the shape of a truncated pyramid. He was a good-natured genie. However, he committed an indiscretion by getting hold of the pyramid's secret. Afterward, he was locked and could not be freed till the time when the human mind would be capable of discovering and implementing the hidden concept in the correct way. In the meantime, however, the imprisoned Genie was permitted to give away some hints which could help men discover the secret.

The next episode took place around 50 AD in Alexandria. At that time, it was the famous Heron of Alexandria who contributed further to bringing the Genie out of its prison. Heron, also called Hero, was a Greek mathematician and engineer who lived in his native city of Alexandria, in Roman Egypt. He is often considered the greatest experimenter of antiquity. His work is representative of the Hellenistic scientific tradition.

---

[19] "The Fisherman and the Genie" in *Tales of the Dervishes*, Idries Shah (The Octagon Press, London, 1993, p. 117).

Heron was trying to find a more practical method for the calculation of the volume of a truncated pyramid. In practice, it would be difficult to use the Egyptian formula because it represented a rather idealistic case. It required the measurement of the vertical height of the truncated pyramid. The vertical height of a truncated pyramid is a simple parameter. Its measurement, however, is difficult to execute. For practical purposes, therefore, an intermediary parameter is needed which would allow to "translate" an idealistic concept into a more practical version.

The "translation" of an ideal into its "earthly" equivalent is always challenging. Such a "translation" very often dilutes the original formula by introducing additional relationships, which may interfere with the original concept. In the case of the truncated pyramid, it was the length of the slant edge of the pyramid that could serve as such an intermediary parameter:

The double arrow indicates the length of the slant edge.

Therefore, Heron opted for the use of a formula that allowed to calculate the vertical height by knowing the length of the slant edge of the pyramid. In the algebraic notation, this formula may be presented as:

$$h = \sqrt{s^2 - \frac{(a-b)^2}{2}}$$

where "s" is the length of the slant edge, "a" and "b" are the lengths of the base and the top squares, respectively.

Heron's method was described in the *Stereometria,* a work published around 100 AD. First, Heron calculated the vertical height of a pyramid with the lower side equal to 10 units, the upper side equal to 2 units, and the slant edge (or the slant height) equal to 9 units. He found that the vertical height is equal to $\sqrt{49} = 7$, which was the correct result. Afterward, he attempted to solve the same problem for another pyramid. He chose a pyramid with the following dimensions: 15 units for the slant edge length, 28 units for the bottom size, and 4 units for the top size.[20] Using his formula, he should have arrived at a solution where the height of the pyramid (h) is equal to:

$$h = \sqrt{(225 - 288)} = \sqrt{-63}$$

But then something strange happened. Somehow, instead of the square root $\sqrt{(225 - 288)}$ required by his formula, Heron reversed the two numbers and took the square root $\sqrt{(288 - 225)}$. He ignored the minus sign under the square root and simply wrote the result as $\sqrt{63}$.

Heron fudged his calculation by dropping the negative sign. In this way, he missed being the earliest known scholar to have derived the square root of a negative number in the mathematical analysis of a physical problem. Heron of Alexandria failed to recognize the hidden property of the problem that he was working on.

---

[20] W.W. Beman, "A Chapter in the History of Mathematics," *Proceedings of the American Association for Advancement of Science,* 46 (1897) 33-50. (Quoted by Paul J. Nahin in *An Imaginary Tale,* Princeton University Press, Princeton, N.J., 1998).

The hidden property of the truncated pyramid is related to the fact that the length of the slant edge and the vertical height are not independent values. There is a specific relationship between them; they cannot be chosen arbitrarily. Yet, Heron ignored that relationship. In Heron's example, the vertical height of the pyramid is longer than the slant edge length: an impossible condition. Heron's pyramid was an abstract structure. Such an *imaginary* pyramid could be perceived conceptually – but it could not be constructed. The imaginary pyramid was an allegorical representation of something that exists outside of the physical dimension.

It is quite surprising that Heron, the greatest experimenter of antiquity, chose such an imaginary case. As a result, he ended up with an equation whose solution required the square root of a negative number. At that time, negative numbers were not accepted yet by mathematicians. The concept of a negative number, i.e., something less than *nothing*, was viewed as a strange and unrealistic idea. Such a concept was beyond the grasp of a man living in the 1st century AD. Not to mention such an oddity as the square root of a negative number. Yet, it was that odd entity that was to revolutionize human cognition. This was the first encounter of the rational mind with the imaginary world. The concept of the imaginary world was hidden within the Egyptian truncated pyramid and waiting for a time when it could become reality.

Heron's calculation is an example that illustrates that when an idea, a concept, or even an object is unknown or unfamiliar – it remains "invisible" to the human mind. As mentioned earlier, negative numbers were an unknown entity at that time. They did not enter yet into the human mind. Therefore, they simply … did not exist. In other words, Heron could not see them. He wrote his result in a form that was familiar to him. Before the mind can accept a new idea, it must be gradually prepared for it. This is why these sorts of ideas and concepts are first introduced in their allegorical forms in various fairy tales.

Historians still do not know why Heron chose such an imaginary example for his calculations. As historians do not believe in genies, they are left in the dark. For us, it is a simple fact. It was the Genie that tricked Heron into working with an imaginary shape of the pyramid. This was the Genie's hint. Heron, however, completely missed that hint. We can imagine all the swearing and cursing that the Genie called upon the head of poor Heron. However, the time was not ripe for our Genie to be freed yet. For the time being, the Genie's secret had to remain hidden. Just like in the tale from the *Arabian Nights*, the bottle was thrown back into the sea:

> …The fisherman was terrified, and, casting himself upon the sand, cried out: "Will you destroy him who gave you your freedom?"
>
> "Indeed I shall," said the genie, "for rebellion is my nature, and destruction is my capacity, although I may have been rendered immobile for several thousand years."
>
> The fisherman now saw that, far from profit from this unwelcoming catch, he was likely to be annihilated for no good reason that he could fathom.
>
> He looked at the seal upon the stopper, and suddenly an idea occurred to him. "You could never have come out of that bottle," he said. "It is too small."
>
> "What! Do you doubt the word of the Master of Jinns?" roared the apparition. And he dissolved himself again into wispy smoke and went back into the bottle. The fisherman took up the stopper and plugged the bottle with it.
>
> Then he threw it back, as far as he could, into the depths of the sea….[21]

---

[21] See Note #19.

***

As the cultural center of gravity shifted from the Roman province of Egypt to the Abbasid Caliphate, so our tale follows it by moving from Alexandria to Bagdad. This should not be a surprise: the modern literary origin of the genies was the Bagdad of Harun al-Rashid, the fifth Caliph of the Abbasid dynasty.

In 750 AD, the Abbasid dynasty replaced the Umayyad as the ruling family of the Islamic Empire. The Abbasids were influenced by Mohamed's saying, "the ink of a scholar is more holy than the blood of a martyr," which emphasized the value of knowledge. The Abbasids championed the cause of knowledge. In 762, al-Mansur, the second Abbasid Caliph, founded Bagdad and made it his capital. Bagdad became the intellectual hub of the world. Al-Mansur built a palace library that became a center of science, culture, and philosophy. The library included living quarters for scientists, philosophers, and translators. The latter had the specific task of rendering into Arabic scientific and philosophical texts of the ancient Greeks. Harun al-Rashid, the fifth Caliph, ruled the Caliphate during the peak of the Islamic Golden Age, from 786 to 809. Bagdad became the new cultural capital of the world, taking over that title from Egypt's Alexandria. Bagdad's library was transformed into the legendary House of Wisdom. The character of Harun al-Rashid was immortalized in many stories included in the *Arabian Nights*.

One of the most prominent guests of the House of Wisdom was the Persian mathematician and astronomer Muhammad ibn Musa al-Khwarizmi, who lived between 780 and 850. As his name indicates, he was a native of Khwarizm (today Khiva), a city of present-day Uzbekistan. The term *algebra* comes from the title of one of his books. The term *algorithm* is the Latinized version of al-Khwarizmi's name (*Algorizmi*). Because he was the first to treat algebra as an independent discipline and introduced the methods of solving first- and second-degree equations, he has been described as the father of algebra. Second-degree equations, also known as quadratic equations, are of particular importance to our story. It was a quadratic equation that held the key needed for getting the Genie out of the bottle.

Khwarizmi was working on several types of quadratic equations. As far as our tale is concerned, the most interesting detail about al-Khwarizmi is the fact that he, just like Heron of Alexandria, persisted in ignoring zero and negative solutions in his techniques of solving the equations. He considered only positive solutions because they are the only solutions that can be interpreted geometrically.

Although enclosed in his bottle, the Genie could exercise some limited powers. Following the rule which says that "A particular level of consciousness may only be fully developed while struggling towards the next higher one," the Genie managed to encourage mathematicians to focus on an even more challenging problem: the cubic equations. The cubic (third-degree) equations are more complicated, and they cannot be solved satisfactorily without the use of negative numbers. And this was the Genie's clever trick because to solve a cubic equation, one must reduce it to its quadratic form. In other words, the technique for solving cubic equations required perfecting the method of dealing with quadratic ones. This is why our tale will now follow the events which led to the discovery of the solution of cubic equations.

As Khwarizmi did not attempt to expand his analysis, then somebody else had to be encouraged to do so. Therefore, our tale has to move somewhere else. The next episode took place in the city of Nishapur at the end of the 11th century AD.

Nishapur was the metropolis of Khorasan, a province in today's northeastern Iran. It was there that Omar Khayaam was born. Khayaam was an accomplished astronomer and mathematician. As a mathematician, he was the first to recognize the importance of cubic equations. He is known as the author of the first theory of cubic equations. He identified fourteen different types of cubic equations. For each of these fourteen types, he provided one of the solutions. He analyzed cases in which a cubic equation had no solutions, one solution, or two solutions.

Khayaam's method was similar to that of the second-degree equations described by al-Khwarizmi. And again, just as al-Khwarizmi, Khayaam considered only positive solutions which he could derive by using geometrical methods. In these methods, the solutions are found as intersections of various geometrical shapes. Only a limited number of cubic equations could be resolved in this way. The remarkable thing about Khayaam was his overall understanding of the situation. He knew that there was something that could not yet be expressed in the language of mathematics. This is reflected by his statement: "Maybe one of those who will come after us will succeed in finding it."[22] It may help to understand this statement to recall that Khayaam was a mystical poet. In the West, he is known as the author of the famous quatrains entitled *The Rubaiyyat*. As a mystic, he was aware of the requirement of the right place, people, and time. He knew that the conditions were not ready yet for a general solution to this particular algebraic problem. Instead, he used his poetry to indicate the overall concept hidden within the cubic equations. In *The Rubaiyyat* he wrote:

> Yet those who proved most perfect of our kind
> Mounted the soaring Burak of their thoughts.
> Study your essence: like the Firmament,
> Your head will turn and turn, vertiginously.[23]

---

[22] *The Secret Formula*, Fabio Toscano; translated by Arturo Sangalli (Princeton University Press, Princeton, NJ, 2020).
[23] *A Journey with Omar Khayaam*, W. Jamroz (Troubadour Publications, Montreal, 2018, p. 50).

In these lines, Khayaam alludes to moments of inspiration. He says allegorically that such flashes of higher consciousness ("the soaring Burak of their thoughts") may be induced by momentarily emptying one's mind from shallow thoughts. Such sorting out of deeper thoughts from shallow ones may be compared to using a centrifugal force ("Your head will turn and turn") to separate lighter substances from heavier ones. In other words, there are qualitatively different levels of consciousness. As long as higher levels of consciousness are not activated, certain concepts and ideas cannot be perceived. Khayaam knew that "Man can use only what he has learned to use." Therefore, he cast the bottle back into the ocean – just like the grandson of the fisherman in the previously quoted episode from "The Fisherman and the Genie":

... Many years passed, until one day another fisherman, grandson of the first, cast his net in the same place, and brought up the self-same bottle.
He placed the bottle upon the sand and was about to open it when a thought struck him. It was the piece of advice that had been passed down to him by his father, from his father.
It was: "Man can use only what he has learned to use." ...
The young fisherman, remembering his ancestral adage, placed the bottle carefully in a cave and scaled the heights of a nearby cliff, seeking the cell of a wise man who lived there.
He told the story to the wise man, who said: "Your adage is perfectly true: and you have to do this thing yourself, though you must know how to do it."
"But what do I have to do?" asked the youth.
"There is something, surely, that you feel you want to do?" said the other.
"What I want to do is to release the Genie, so that he can give me miraculous knowledge: perhaps mountains of gold, and seas made from emeralds, and all the other things which genies

can bestow."

"It has not, of course, occurred to you," said the sage. "that the Genie might not give you these things when released; or that he may give them to you and then take them away because you have no means to guard them; quite apart from what might befall you if and when you did have such things, since 'Man can use only what he has learned to use'."

"Then what should I do?"

"Seek from the Genie a sample of what he can offer. Seek a means of safeguarding that sample and testing it. Seek knowledge, not possessions, for possessions without knowledge are useless, and that is the cause of all our distractions."

Now, because he was alert and reflective, the young man worked out his plan on the way back to the cave where he had left the Genie.

He tapped on the bottle, and the Genie's voice answered, tinny through the metal, but still terrible: "In the name of Solomon the Mighty, upon whom be peace, release me, son of Adam!"

"I don't believe that you are who you say and that you have the powers which you claim," answered the youth.

"Don't believe me! Do you not know that I am incapable of telling a lie?" the Genie roared back.

"No, I do not," said the fisherman.

"Then how can I convince you?"

"By giving me a demonstration. Can you exercise any powers through the wall of the bottle?"

"Yes," admitted the Genie, "but I cannot release myself through these powers."

"Very well, then: give me the ability to know the truth of the problem which is on my mind."

Instantly, as the Genie exercised his strange craft, the fisherman became aware of the source of the adage handed down by his grandfather. He saw, too, the whole scene of the release of the Genie from the bottle; and he also saw how he could convey to others how to gain such capacities from the genies. But he also realized that there was no more that he could do. And so the fisherman picked up the bottle and, like his grandfather, cast it into the ocean.

And he spent the rest of his life not as a fisherman but as a man who tried to explain to others the perils of "Man trying to use what he has not learned to use." ...[24]

After Khayaam, many mathematicians tried to find a solution to the mystery of third-degree equations. Yet, the general formula for the cubic equations remained an unsolved problem. Our Genie was still locked in his bottle.

Some four hundred years later, the eminent Luca Pacioli (1447 – 1517), a Franciscan friar and a famous Italian mathematician pronounced a verdict on the cubic equations that influenced several generations of mathematicians. Pacioli wrote: "It is not possible to solve the general form of cubic equations by means of an algebraic formula with the algorithmic tools available." We can imagine the Genie's desperation when he heard this.

We may notice the difference between Pacioli's view and that expressed by Khayaam. Khayaam, as a mystic, was able to perceive the way the human mind evolves. Pacioli was not a mystic; he was a craftsman.

At the beginning of the 16th century, the general formula for cubic equations remained an impenetrable mystery.

---

[24] See Note #19.

Italian Duels

But people only die in proper duels, you know, with real wizards.
(J.K. Rowling)

In the Middle Ages, trial by combat was a method prescribed by Germanic law to settle accusations in the absence of witnesses or of a confession. In other words, the resolution of a dispute between two parties was delegated to a fight in single combat. The winner of the duel was proclaimed to be right. In essence, it was a judicially sanctioned combat. It remained in use throughout the European Middle Ages.

A duel was also fought as a result of an insult or challenge to one party's honor which could not be resolved by a court. Weapons used in duels were standardized and typical of a knight's armory, for example, longswords, axes, polearms. The choice of weapon was at the discretion of the challenged knight. A duel lasted until one party could no longer fight back. In early cases, the defeated party was then executed. This type of duel soon evolved into the more *chivalric pas d'armes* or "passage of arms," which became fashionable in the late 14th century and remained popular through the 15th century. A knight or group of knights would stake out a traveled spot, such as a bridge or a city gate. It would then be announced that any other knight who wished to pass must first fight or be disgraced. If a traveling knight did not have weapons or horse to meet the challenge, he would be provided with it. If a knight would choose not to fight, he would leave his spurs behind. This would be treated as a sign of utmost humiliation. If it happened that an unescorted lady would pass that way, she would have to leave behind a token such as a glove or a scarf. This token would have to be rescued by a knight who passed there sometime later on. Only after a fight could the winning knight return the token to the lady.

During the early Renaissance, dueling established the status of a respectable gentleman and was an accepted manner to resolve disputes. The first published "code of dueling," appeared in Renaissance Italy.

By the 17th century, dueling had become regarded as a prerogative of the aristocracy. Attempts to discourage or suppress dueling generally failed. For example, King Louis XIII of France outlawed dueling in 1626. His successor Louis XIV intensified efforts to wipe out the duel. Despite these efforts, dueling continued unabated, and it is estimated that between 1685 and 1716, French officers fought 10,000 duels, leading to over 400 deaths.

By about 1770, duels underwent a number of important changes in England. Firstly, unlike their counterparts in many continental nations, English duelists enthusiastically adopted the pistol. Special sets of dueling pistols were crafted for the wealthiest of noblemen. Also, the office of "seconds" was introduced. The "seconds" or "friends" were nominated by the aggrieved parties to conduct their dispute. The "seconds" would attempt to resolve the dispute upon terms acceptable to both parties. When they failed, they would arrange and oversee the means and the conditions of a duel.

In Russia, dueling was first discovered by the officer corps in the early 1700s. Russian officers took to it with such enthusiasm that the Tsar had to forbid the practice for fear that there would soon be no one left to lead his troops. Here is an account of dueling in Russia, given by Count Rostov in *A Gentleman in Moscow*, a novel written by Amor Towles:

Why is it that our nation above all others embraced the duel so wholeheartedly? he asked. ... Some, no doubt, would simply dismiss it as a by-product of barbarism. Given Russia's long, heartless winters, its familiarity with famine, its rough sense of justice, and so on, it was perfectly natural for its gentry to adopt an act of definitive violence as the means of resolving disputes. But in the Count's considered opinion, the reason that dueling prevailed among Russian gentlemen stemmed from nothing more than their passion for the glorious and grandiose. True, duels were fought by convention at dawn in isolated locations to ensure the privacy of the gentlemen involved. But were they fought behind ash heaps or in scrapyards? Of course not! They were fought in a clearing among the birch trees with a dusting of snow. Or on the banks of a winding rivulet. Or at the edge of a family estate where the breezes shake the blossoms from the trees. ... That is, they were fought in settings that one might have expected to see in the second act of an opera. In Russia, whatever the endeavor, if the setting is glorious and the tenor grandiose, it will have its adherents. In fact, over the years, as the locations for duels became more picturesque and the pistols more finely manufactured, the best-bred men proved willing to defend their honor over lesser and lesser offenses. So while dueling may have begun as a response to high crimes –to treachery, treason, and adultery– by 1900 it had tiptoed down the stairs of reason until duels were being fought over the tilt of a hat, the duration of a glance, or the placement of a comma. In the old and well-established code of dueling, it is understood that the number of paces the offender and offended take before shooting should be in inverse proportion to the magnitude of the insult. That is, the most reprehensible affront should be resolved by a duel of the fewest paces, to ensure that one of the two men will not leave the field of honor alive.[25]

At this point, you may be wondering what dueling has to do with our Genie and the problem of the cubic equations?

It turns out that dueling was a key factor leading to the creation of the right "ambiance" for the freeing of the Genie from its bottle. It was in a dueling ambiance that the next episode of our tale took place. It was in northern Italy in the 16th century.

***

In parallel to trial by combat, another type of dueling was instituted in Renaissance Italy. This particular combat involved …. famous mathematicians.

Duels of mathematicians were very much in vogue in Italy in those days. Any scholar could be approached by an amateur mathematician or anyone wishing to challenge him. And this was not just a manner of speaking. The duels in which mathematicians challenged one another were carried out in a way reminiscent of chivalry tournaments. Just like in chivalric combat, a mathematician or a scholar would send to his rival a "challenge gauntlet." The challenge consisted of a list of problems to be solved in a given period of time. After that specified time, the recipient of the challenge would propose a set of his own problems to his rival. An unwritten rule of Renaissance mathematical duels was that a challenger could not propose to his rival any problem he was not

---

[25] *A Gentlemen in Moscow*, Amor Towles (Penguin Books, New York, 2018.)

able to solve himself. For that reason, it was common for mathematicians to keep their discoveries secret with the purpose of using them as secret weapons when they were challenged by others. In other words, new mathematical discoveries served as secret personal weapons to defend their author's position and status. Very often, the discovery would never be disclosed; their discoverers would take them to their graves.

Mathematical duels had a long history and some illustrious predecessors. For example, during the first half of the 13th century, several mathematical duels had as protagonist Leonardo Fibonacci of Pisa. Fibonacci was the most talented Western mathematician of the Middle Ages. He became famous for his discovery of a sequence of numbers known as Fibonacci numbers. Fibonacci attracted the attention of the Holy Roman Emperor Frederick II – a renowned patron of sciences. In 1225 Frederick visited Pisa and held a mathematical duel to test Fibonacci's talents. Several well-known mathematicians were invited to take part in the competition. Johannes of Palermo, a mathematician working for the emperor, presented three questions to Fibonacci and his challengers. Fibonacci solved them all while none of the other mathematicians was able to solve any. They all withdrew without providing a single solution. Later on, Fibonacci wrote *Liber quadratorum*, a book in which he recounted the problems he was challenged with in the duel. A revised edition of the book was dedicated to his friend Michael Scott, an astrologer at Frederick's court.

Tradition required that, in case of disagreement, a public debate should be held in which the contenders would discuss the disputed problems and solutions. This would take place in front of judges, notaries, government officials, and a large crowd of spectators. It was not unusual in those duels for tempers to flare, and personal abuse to take the place of scientific argument. Admittedly, the stakes could be very high: the winner of a public mathematical duel, whoever had solved the largest number of problems, gained glory and prestige. Often there would also be a monetary prize, enrolment of new fee-paying disciples, an appointment or confirmation to a university chair, a salary increase, and maybe well-paid professional commissions. The defeated contender's future career, on the other hand, risked being seriously compromised.

***

The next episode of our tale involves Gerolamo Cardano. Cardano was born in Pavia, Lombardy, in 1501. He was a polymath with a wide range of interests. He invented the combination lock, the gimbal consisting of three concentric rings used to support the compass or gyroscope, the Cardan shaft with universal joints, and many other devices. However, his biggest achievement was related to modern algebra.

Like most of the heroes of our tale, Gerolamo Cardano was not an ordinary man. First of all, modesty was not part of his repertoire:

> Wherever I happened to be, in Milan, Pavia, Bologna, France, or Germany, since I was twenty-three, I never met anyone who could measure up to me in a discussion or debate, but I'm not boasting about that. If I had been a stone, the result would have probably been the same, because this is not a privilege of my nature or its greatness, but it's due to the ignorance of those who challenge me.[26]

---

[26] *The Secret Formula*, Fabio Toscano – see Note #22 (The book gives a detailed description of Cardano's life and interactions with Tartaglia. It also includes extensive documentation and bibliographic data).

His father left him a rather meager inheritance. While studying at the University of Padua, he could hardly earn enough money to pay for his expenses. To make ends meet, Gerolamo resorted to his passion: games of chance. Cards, dice, and chess became his prime source of income. This allowed him to finish his study and obtain a doctorate in medicine. Later on, he even wrote a book entitled *Book on Games of Chance*. The book contains the first systematic treatment of probability. It was based on his experiences with the game of dice. He was able to demonstrate the efficiency of defining odds as the ratio of favorable outcomes to the total number of possible outcomes. The book also contains a section on cheating methods.

Since his youth, Gerolamo believed that he possessed extraordinary powers allowing him to predict the future through dreams. He claimed that he could hear a humming sound every time someone was talking about him. He was also convinced that his illuminating power protected him from enemies and imminent danger. He was fond of magic, horoscopes, divination, and occultism. He carried amulets and precious stones with mystical powers to guard himself against harmful influences. No wonder that some of his contemporaries thought that Gerolamo was not in his right mind. Despite all of these and thanks to his influential friends, in 1539 Gerolamo was admitted as a member of Millan's College of Physicians. Soon he established himself and became Milan's most famous and sought-after physician.

At the same time, Gerolamo was pursuing his interest in mathematics. He began to write a book on applied algebra. Regarding the theory of equations, Gerolamo accepted Pacioli's claim of the impossibility of finding a general formula to solve the cubic equations. Therefore, he decided to exclude them from his book and limit himself to the theory of first and second-degree equations. It was during that time that Gerolamo was visited by Antonio Maria Fior. Fior challenged him with some problems involving third and fourth-degree equations. Then Fior told him that, notwithstanding Pacioli's claim, someone certainly possessed a secret formula for cubic equations. He also told him about the sensational Venetian challenge in which he himself took part a few years earlier.

The Venetian duel was between Fior and Niccolo Tartaglia. (Tartaglia means "The Stammerer;" he was called by this name because of his difficulty in speaking after he received a sword cut to the tongue during the French siege of Brescia, the city where he was born.) Fior proposed thirty problems to Tartaglia, who in the response sent the challenger thirty of his questions. As was customary, the winner would be whoever answered the most questions. On February 22, 1535, Fior and Tartaglia entrusted their respective lists of questions with a notary in Venice, agreeing to hand over the solutions fifty days later. What were the stakes of the contest? First of all, honor and reputation, and then a lavish dinner at a tavern for each unsolved problem, with the bill to be footed by the unfortunate contestant who capitulated before his rival's question.

In the ensuing months, echoes of the Venetian challenge spread well beyond the city, carrying the news of its resounding outcome. Tartaglia had literary humiliated Fior by solving in a couple of hours all thirty problems posed by his opponent, while the latter had not been able to answer a single one of Tartaglia's questions. The most astonishing fact of that duel was the topic on which Fior challenged his opponent. All his questions were related to cubic equations. Fior had believed that Tartaglia could not solve any of them because Pacioli had stated that their solution was impossible. By sheer luck, Tartaglia had found the answers to Fior's questions just eight days before the date of the submission. He had discovered the formula for the solution of one type of third-degree equations.

Let us recall that the mathematical duels obeyed a gentleman's agreement, which required that a contender could not propose a problem if he did not know how to solve it himself. If Fior challenged his opponent with problems involving these particular cubic equations, then he himself must have known the solution. Indeed, sometimes later on, Fior disclosed that thirty years earlier an unnamed mathematician had revealed to him the secret.[27]

The news about the existence of a solution for cubic equations galvanized Cardano into action. He probably had tried to find the formula by himself, but without success. With his mathematical book soon going to press, Gerolamo decided to contact the discoverer of the formula directly. In January 1539, he asked a friend to carry a message to Tartaglia. He offered to include Tartaglia's solution in his book and give him full credit. Tartaglia refused. Afterward, Cardano and Tartaglia exchanged several messages that led to their meeting in Milan in March of the same year. During the meeting, Tartaglia disclosed his secret formula under the condition that Cardano would take an oath of secrecy and not include it in his book. Cardano took the oath. Once the formula was revealed, the meeting came to an end. In May 1539, Cardano sent a freshly printed copy of his book to Tartaglia. Cardano kept his word. Cubic equations were not mentioned and no reference was made to Tartaglia's formula.

Afterward, Cardano continued his work on the analysis of cubic equations. By analyzing Tartaglia's solution, Cardano found its limitation. The nature of this limitation was the same as that encountered by Heron, Khwarizmi, and Khayaam. For certain forms of the equations, the solutions seemed to require negative numbers. At that time, Cardano reported the problem to Tartaglia. Tartaglia, however, was unable to explain it.

In 1542, Cardano visited a friend in Bologna. His host showed him an old notebook that had belonged to Scipione Dal Ferro – his father-in-law who had died sixteen years earlier. When Gerolamo

---

[27] *Ibid.*

opened the notebook, he found in it the formula for the solution of a certain type of cubic equations which Dal Ferro had discovered some twenty years earlier. It was the same formula that was discovered independently by Tartaglia. Dal Ferro is a person shrouded in mystery and little is known about him. By keeping his formula secret, Dal Ferro merely conformed to the practice of his day, saving it for use in eventual duels against challengers. He disclosed it only to a few friends, his son-in-law and ... Fior. This is how Fior knew the formula when he challenged Tartaglia.

After returning from Bologna to Milan, Cardano felt released from any obligation from his oath to Tartaglia. As far as he was concerned, the time for the secret was over. In 1545 Cardano published his book entitled *Ars Magna*, in which he presented solution rules for third-degree equations. In the book, he identified the names of the authors, i.e., Dal Ferro and Tartaglia, and gave them the credit they deserved.

In *Ars Magna*, however, Cardano described problems that indicate that he found the ultimate solution for cubic equations. It was a solution that was not known either to Dal Ferro or Tartaglia. Cardano's solution represented quite a radical approach to algebraic equations. Cardano applied his solution to a problem which was expressed in a single sentence:

Divide 10 into two parts in such a way that their product shall be 40.

This may be described in the following way. We are dealing with a problem that is set in the real world. Both numbers "10" and "40" are as real as real numbers can be. Yet, it is impossible to find two real numbers which would satisfy both conditions, i.e., that their sum is 10 and their product is 40. Yet, Cardano found a solution. How did Cardano find it? It is then and there that our Genie, in a slightly different shape, reappears:

One evening, when Cardano was working on this particular problem, a stranger showed up at the door of his house. The stranger was impeccably dressed and displayed aristocratic manners. Without introducing himself, he said:
"You are seeking a certain knowledge. You are working on a solution for an old problem. I came here to help you. But first, you must answer this question:
Do you desire to learn why you are not able to solve it or do you just want to get the solution?"
"All I want is to know the solution for this problem!" – answered Cardano without any hesitation. "If I know the solution then I, by myself, will be able to figure out what I was missing" – he added with a smile.
"In such a case," said the stranger, "you do not need me. Instead, this crystal will help you."
He took out of his pocket a shiny yellowish stone in the shape of a truncated pyramid.
"Take this stone. When you rub it, you will be provided with the solution you want. But you may rob the stone only twice. Afterward, the stone will be useless."
After delivering his short message, the stranger left.
Cardano did not wait.
He rubbed the stone and an apparition appeared. It was in the shape of an imp, a small and ugly imp, with pointed ears.
"I am here, master, I am here!" he cried, "Your slave is at your service. Ask what you will, for I am the slave of him who holds my master's stone."
Cardano was used to various strange and mysterious happenings. Therefore, he was not taken aback at all by what

was happening. All he wanted to know was the solution to his problem.

"Show me how to solve this problem" – he demanded.

"You have to understand," the apparition started to explain, "that these *two parts* you are looking for do not belong to this world. They belong to an invisible, imaginary world. You may see them when you rub the stone again. They will appear in their shapes on the piece of paper that is in front of you. You will have to write down my name at the end of each of them. My name is *Imp*. Afterward, you may check the validity of the solution by applying the conditions prescribed in your problem. Just remember that when you do square my name, it will be transformed into a new one, which in your language of signs means (–1)." Saying that the apparition disappeared. Cardano did as he was told. As soon as he rubbed the stone again, he saw two terms appearing on the piece of paper. They looked like rather ordinary and simple algebraic expressions:

$$5 + \sqrt{15} \quad \text{and} \quad 5 - \sqrt{15}$$

He quickly wrote down the name of *Imp* at the end of each of these terms:

$$5 + \sqrt{15} \, Imp \quad \text{and} \quad 5 - \sqrt{15} \, Imp$$

With great excitement, he started to apply the conditions prescribed in the problem. He was astonished to find out that, as required by the original conditions, their sum was equal to 10. At the same time, *Imp's* names disappeared!

$$5 + 5 = 10$$

Then he multiplied these two terms as required by the original prescription and applied Imp's instruction. This was even more surprising! Like in the first case, the names of *Imp* also disappeared. And he obtained the correct result:

$$25 + 15 = 40$$

A satisfying smile lightened Cardano's face. "Finally I got it!" he said to himself.

Cardano did not include the name of *Imp* in his book. Neither did he explain how he derived his solution. He just copied the solution by using negative numbers. He referred to the negative numbers under the square roots as *refined* and *sophistic*. But it seems that he had not understood them completely. It was left up to the next generations of mathematicians to work out their fuller meanings and applications. Later, these numbers started to be referred to as *imaginary* numbers. The sum of a real number (a) and an imaginary number (b) became known as a *complex* number (z). Coincidently, mathematicians selected the letter i to denote the imaginary part of a complex number:

$$z = a + ib$$

In this way, *Imp's* initial has been permanently imprinted in the texts and books of mathematics.

Cardano's solution forced mathematicians to take both negative and imaginary numbers seriously. It should be added, however, that certain anxiety and hesitance against imaginary numbers lasted well into the 19th century. In other words, complex numbers were a solution waiting for a problem. The problem for which they were needed appeared only in the early 20th century.

At the end of his life, Cardano got into serious trouble for some of his odd activities. In his writings, he inserted some obscure passages such as casting a horoscope for Jesus and a tribute to Nero, the tyrannical Roman Emperor. Consequently, he was accused of heresy by the Holy Office. He was sentenced by the Inquisition to several months in prison followed by house arrest. Cardano died in Rome in 1576.

***

Cardano's solution could be expressed only using complex numbers. Although his solution played a crucial role in the formation of modern algebra, its importance impacted a much more general area of human cognition. This was an example of how, by using an imaginary element, it was possible to solve things belonging to the physical world. It was the first time that such a possibility entered the consciousness of the rational mind.

Cardano's algebraic solution may be paraphrased into a narrative, a sort of fairy tale. Such a tale would run like this:

> Two events occurred in the physical world. They were observed and recorded. Symbolically, these events were described by two numbers, 10 and 40. It was known that these two events were related to each other. Furthermore, these two events had a common origin; they were the outcome of two previous events. These previous events were described as:
>
> $$5 + \sqrt{15}i \quad \text{and} \quad 5 - \sqrt{15}i$$

This means that for events 10 and 40 to take place and be related, they would have to have their origin in an *imaginary world*.

Although in this particular case the problem is limited to algebra, the provided solution is an allegorical illustration of the working of the cosmic matrix. Namely, it illustrates that the original template of certain events and relationships is placed within the "imaginary" or the invisible world. This means that these types of relationships may be fully understood only by looking at them from the macrocosmic point of view. Therefore, to grasp their meanings, one must analyze them from the perspective of the cosmic matrix.

Cardano's solution was a symbolic manifestation of the relationship between two different zones of consciousness, physical and macrocosmic. The interesting thing is that this concept was implanted in the human mind at the time of the construction of the Egyptian pyramids. It was symbolically indicated by the missing capstone. There was nothing new needed to accept it. The acceptance of such a concept did not require a new discovery, a new experiment, or new equipment. Neither did it have to be derived from previously unknown physical laws or mathematical rules. Everything that was needed was there from the very beginning. The only obstacle was a certain block in the human mind. This concept was beyond the grasp of ordinary human rationale. Let us recall that, during all that time, the concept in its allegorical form was on display through certain tales, such as some biblical stories and the tales of the *Arabian Nights*. It looks like it took nearly three millennia before this concept could be correctly assimilated in the human mind.

Cardano's solution allows us to take a more accurate look at the term "genius." It becomes obvious that there are two types of geniuses. The first type is incredibly talented and skilled individuals who can comprehend the most sophisticated things; they can understand them, explain their mechanism, and extend their application to a range of other problems. They can work at the limits of intellectual rationale. They are the *craftsmen*. The other type is like *magicians*. They are assisted by a genie. Therefore, the working of their minds is incomprehensible to rational humans. It is those magicians that, from time to time, are needed to advance human imagination. Cardano was one of them.

Cardano's *magical* thinking allowed him to remove a certain veil that was blocking the human mind from perceiving a concept that was needed for furthering the evolution of human consciousness. Coincidently, this development took place after a major change occurred within the structure of the Macrocosm some three hundred years earlier. It was also at that earlier time that the second version of the *Arabian Nights* was published. This newer version of the *Arabian Nights* contained, among other things, an allegorical form of the solution that Cardano used for the cubic equations.

The Second Arabian Nights

The cosmic religious experience is the strongest and the noblest driving force behind scientific research.
(*Albert Einstein*)

After Mohammed's Night Journey, the next account of the structure of the Macrocosm was given by Ibn Al-Arabi. Al-Arabi was an Andalusian scholar, poet, and philosopher. In his book entitled *The Meccan Revelations*, he described his own "nocturnal journey" into the Macrocosm.

Al-Arabi's "journey" took place some six hundred years after Mohammed's Night Journey. Yet, the structure of the Macrocosm described by Ibn Al-Arabi in *The Meccan Revelations* was identical to that experienced by Mohammed. In Ibn-Arabi's description, there were neither traces of Mohammed's presence nor Mohammed's effect on the structure. In other words, this description was a copy of that reported by Mohammed. This would imply that there had been no change within the Macrocosm during that period. This seems to be rather unlikely. Something must be missing from Ibn Arabi's account.

Later on, Ibn-Arabi gave another account of the macrocosmic structure. He described a second experience in the book entitled *The Bezels of Wisdom*. He wrote this book in the later years of his life, i.e., near the year 1230. It was then that he augmented his previous report.

In *The Bezels of Wisdom*, Ibn-Arabi explained the circumstances that led him to the writing of that book. Namely, *The Bezels of Wisdom* was passed to him in a vision. In this vision, he saw Mohammed who "had in his hand a book and he said to me, 'This is the book of the bezels of Wisdom; take it and bring it to men that they might benefit from it'." Ibn Arabi added:

> I therefore carried out the wish, made pure my intention, and devoted my purpose to the publishing of this book ... without any addition or subtraction. ... It is utterly free from all the purposes of the lower soul, which are ever prone to deceive.[28]

The Bezels of Wisdom breaks quite substantially from previous works of Ibn Al-Arabi. Unlike in his other works, Ibn Al-Arabi did not arrange the chapters of the book in any previously used pattern, whether chronological, numerological, astrological, alchemical, or cosmological. It seems that Ibn-Arabi's later experience was of quite a different nature than his previous ones. This later experience was more intuitive. It was free from the intellectual and interpretative renderings which were so typical in Ibn-Arabi's previous works. Just as he said, the book "is utterly free from all the purposes of the lower soul."

Each chapter of *The Bezels of Wisdom* is dedicated to a prophet and to one aspect of wisdom. For example, the first chapter is entitled "The Wisdom of Divinity in the Word of Adam" and the last chapter bears the title "The Wisdom of Singularity in the Word of Mohammed." Among the names figuring in the titles of the chapters are Mohammed's and all the prophets' that appeared in Mohammed's Night Journey. In addition, there are other prophets and patriarchs of the *Bible* and several messengers of ancient Arabic

---

[28] *The Bezels of Wisdom*, Ibn Al-Arabi; translation by R.W.J. Austin (Paulist Press, Mahwah, NJ, p. 45).

and non-Arabic nations. All the prophets who appear in *The Bezels of Wisdom* are referred to in the *Koran*. Yet, they do not account for all the prophets who are mentioned in the *Koran*.

The Bezels of Wisdom is divided into twenty-seven chapters. Although the first chapter is dedicated to Adam and the last chapter to Mohammed, the prophets who appear in-between are not arranged according to any known pattern. This would indicate that the number "twenty-seven" serves as a specific symbol underlying a new feature of the macrocosmic structure. The New Cosmos went through a qualitative adjustment. At the time of Ibn-Arabi, instead of a structure of "seven-heavens," the "quality" of the macrocosmic structure was described by twenty-seven words, or aspects, of wisdom. A new qualitatively different "constellation" was in place.

The "words" are the main elements of the macrocosmic structure that is described in *The Bezels of Wisdom*. The term "word," in its singular form, appears in the title of each chapter. Just like in "at the beginning was the word," each "word" is related to the activation of one of the macrocosmic aspects of wisdom.

The "word" is made of letters of the alphabet. The Arabic alphabet consists of twenty-eight letters. Yet, *The Bezels of Wisdom* consists of only twenty-seven chapters. It seems that one chapter is intentionally missing. Of course, the missing chapter is … inexpressible. It is related to *Alif* – the first letter of the Arabic alphabet. *Alif*, the equivalent to the Semitic letter *Aleph*, the Greek letter *Alpha*, and the Latin letter *A*, is often used to symbolize the Absolute. This is nicely illustrated in the following quatrain from Omar Khayaam's *Rubaiyyat*:

> My heart complained: 'I long for inspiration,
> I long for wisdom, to be taught and learn.'
> I breathed the letter A. My heart replied:
> 'A is enough to occupy this house.'[29]

Although "invisible," *Alif* is manifested by the presence of all the remaining letters of the alphabet. In this symbolic representation, *Alif* is considered as containing all the letters. It provides the overall structure within which they are all displayed.

So, what does it mean?

Just like *Alif* contains the overall structure of the alphabet, so the missing chapter serves as the overall frame for all the twenty-seven chapters of *The Bezels of Wisdom*. The overall frame is there - but it is hidden. Therefore, all twenty-seven chapters are contained within that invisible chapter. The invisible chapter provides a link to the macrocosmic matrix.

At the beginning, the Macrocosm was empty. At the time of Mohammed, the macrocosmic structure consisted of seven layers – symbolically referred to as seven heavens. At the time of Ibn-Arabi, the main-frame of the structure was not changed. Instead, it was enriched. *The Bezels of Wisdom* describes the newly activated set of "qualities." Just like the letters of the alphabet, these various "words" or "wisdoms" represent the spectrum of macrocosmic qualities. And this was the message that was encapsulated in Ibn-Arabi's experience. At that time, the Macrocosm contained the set of qualities that had been activated by the "returning" earthly minds. These perfected minds are symbolically represented by the prophets and patriarchs who "returned" to the Macrocosm.

---

[29] *A Journey with Omar Khayaam*, W. Jamroz, p. 77 (see Note #23).

In summary, Ibn-Arabi's experience was a "snapshot" of the updated structure of the New Cosmos which was in place at the beginning of the 13th century.

This "snapshot" is of great interest to us. When this "snapshot" was projected back onto the physical world, it triggered within the minds of ordinary men a series of new concepts and ideas. At first, these new ideas were picked up by mystical poets and writers. Afterward, they influenced the way humanity was evolving. Among other things, these ideas helped to guide the scientists in their effort of formulating the physical model of the universe.

***

Following the changes within the Macrocosm, some qualitative adjustments within the minds of ordinary men were needed. Consequently, some subtler impacts started to appear in poetry and literature. A new character appeared. Let us recall that the genies' function was to indicate the presence of another realm, the existence of invisible worlds. Now, in addition to genies, additional and more subtle impacts were needed to help ordinary men to start to perceive some features of the invisible realm. It was not enough to indicate the invisible world. It was time to expose the human mind to certain qualities of that world. Such a function was delegated to a … *beloved*. It was a *beloved* whose presence strongly affected poems and literary narratives.

There was an early indication of such a *lady* in the *Arabian Nights*. It was encapsulated in one sentence in the story of "Khalifah the Fisherman." It is the episode where Khalifah meets a beautiful slave girl and, as the result of this encounter, "the dark veil of ignorance was lifted from his eyes and he became a wiser man":

Such was the influence of Kut-al-Kulub's words on Khalifah that a new world seemed to unfold before him. The dark veil of ignorance was lifted from his eyes and he became a wiser man.[30]

Obviously, this was not an ordinary beloved. Consequently, this led to the appearance of a new kind of lover. This lover was not obsessed with his sentimental, emotional, or sensual desires. Instead, he was driven by much more powerful and sophisticated forces. These forces could not be explained in terms that are usually associated with ordinary romantic love. While trying to win the approval of his beloved, the new lover had to go through quite an elaborate set of experiences. If he was incapable of passing an initial test, he would be flatly rejected. Here is an example of such a test:

A man once met a beautiful woman. He revealed his love to her.
She said, "Beside me is one who is more beautiful than I, and more perfect in beauty. She is my sister."
He looked to see this woman.
Then the first one said:
"Boaster! When I saw you from afar, I thought that you were a wise man.
When you came near, I thought that you were a lover.
Now I know that you are neither."[31]

---

[30] *Tales from the Thousand and One Nights*, N.J. Dawood, p. 325 (see Note #12).
[31] *The Sufis*, Idries Shah (The Octagon Press, London, 1977, p. 281).

Even after gaining initial approval, a lover would have to go through a set of challenging experiences. Such experiences led to a qualitative change in his raw human nature and he would go through quite a remarkable transformation.

The source of this "new" love was placed outside of the physical realm. The forces associated with that source were able to transmute the ordinary human mind in accordance with the new structure that was being projected from the Macrocosm.

***

Jalaluddin Rumi, a 13th century Persian poet, brought the new symbolic literary form of love to its ultimate perfection. Rumi's masterpiece entitled *Mathnawi* was written thirty years after the publication of Ibn-Arabi's *Bezels of Wisdom*. *Mathnawi* illustrates the effect of man's exposure to love.

Rumi's *Mathnawi* is a complex work. It consists of over twenty-five thousand verses. It is divided into six books. Each book contains between eight and nineteen stories; each story is divided into sections. These sections vary in length; the shortest being only two verses long, the longest well over a hundred verses.

The stories inserted in *Mathnawi* describe experiences that are related to the manifestations of the various qualities that were described by Ibn-Arabi. Rumi explains that these macrocosmic qualities may be acquired by ordinary man through the activation of the latent layers of his mind. In other words, the macrocosmic structure contains the patterns of "perfected" minds. By emulating these patterns, man can gain access to macrocosmic "secrets." The appearance of *Mathnawi* marked the next stage of man's evolution.

In this context, we may look at *Mathnawi* as a new version of the *Arabian Nights*.

Let us take a closer look at the design of *Mathnawi*.

All six books of *Mathnawi* form a coherent narrative. The leading character is a lover. He appears under different names, in different places, and in different environments. Sometimes the lover is a king or a prince, on other occasions, he is a slave or a beggar. For example, in the first story of Book 1, the lover simply has bought a slave-girl who then became his beloved. In this way, Rumi indicates the initial state of the lover, i.e., the state of an ordinary man who assumes that he may "purchase" his happiness.

In the middle of the narrative, the lover goes through a trial. He must pass this trial before he can enter the next stage of his "journey." The lover, however, fails the trial. He is rejected by his beloved. He is still not ready for her. This is illustrated in the story entitled "The Lover and his Mistress who rejected him":

> A certain youth was madly in love with a woman. Fortune, however, did not give him an opportunity to meet her. Whenever he sent a messenger to the woman, the messenger would steal his letter. And if he sent his secretary to deliver the letter, the secretary would change the content of the letter while reading it to her. For seven years that youth was pursuing his quest.
> The lover had no possibility of seeing even his Mistress's shadow. He was only hearing descriptions of her. Except for one brief meeting, during which his heart became enravished with love. Afterward, however much effort he made, that cruel lady would give him no opportunity for another encounter. One night, as he was wandering through the city, he got scared by an approaching patrol. In fear, he ran into an orchard. In

the orchard, he saw his beloved, who was searching with a
lantern for a lost ring. The moment the lover found himself
alone with his mistress, he attempted to embrace and kiss her.
But his mistress pushed him away, saying,
"Do not behave so boldly, be mindful of good manners!"
He said, "Why, there is none present here. None is moving
here but the wind. Who is present? Who will stop me from this
conquest?"
"O madman," said she, "you are a fool! You have not learned
from the wise. You are saying that the wind is moving. Tell me
then, who drives the wind along? The wind is blowing and that
shows that the Mover of the wind is also present."
The lover replied, "It may be I am lacking in good manners,
but I am not lacking inconstancy and fidelity towards you."
His mistress replied, "One must judge the hidden by the
manifest. I see for myself that your outward behavior is bad,
and thence I cannot but infer that your boast of hidden virtues
is not warranted by actual facts. You are ashamed to
misconduct yourself in the sight of men but have no scruple to
do so in the presence of the Mover of the wind, and hence I
doubt your supposed virtues."
The lover then proceeded to excuse himself with the plea that
he had wished to test his mistress and ascertain for himself
whether she was a modest woman. He said that he of course
knew beforehand that she would prove to be a modest woman,
but still, he wished to have proof of it.
His mistress reproved him for trying to deceive her with such
false pretenses. She told him that, after being found at fault, his
only proper course should be to confess. Moreover, she added
that any attempt to put her to a test would have been an
extremely unworthy approach.[32]

---

[32] *The Mathnawi of Jalaluddin Rumi*, Edited and Translated by Reynold A. Nicholson (The E.J.W. Gibb Memorial Trust, 1982, Book IV, 1-387).

This story illustrates an intermediate stage. Accordingly, this story is inserted at the end of Book 3 and it is finished at the beginning of Book 4. Yet, the state of the rejected lover is higher than that from the first story. This means, that some progress has been made. But there is still much work to be done before the lover may arrive at his final destination. And the final destination may not be exactly like what the lover was initially expecting. Here is the story of "The Lover and his Mistress." This story alludes to the kind of challenges that are on the lover's path:

A certain lover recounted to his mistress all the services he had done, and all the toils he had undergone for her sake. He said, "For your sake, I did such and such, in this war I suffered wounds from arrows and spears. My wealth, strength, and fame are gone. On account of my love for you, many a misfortune has befallen me. No dawn found me asleep or laughing; no evening found me with money or means to live." What he had tasted of bitterness and sorrow he was recounting to her in detail, point by point. Not for the sake of reproach. No. He was displaying a hundred testimonies of the trueness of his love. For men of the reason a single indication is enough, but how should the longing of lovers be removed thereby? The lover repeats his tale unworriedly: how should a fish be satisfied with a mere indication and refrain from the limpid water? There was a fire in him. He did not know what it was, but on account of its heat, he was weeping like a candle. The beloved said,
"You have done all this, yet open your ear wide and apprehend well. For you have not done what is the root of the root of love and duty. What you have done are only the branches."
The lover said to her,
"Tell me, what is that root?"
She said, "The root thereof is to die and become nothing. You

have done all else, but you have not died, you are still living. You have to die if you are a self-sacrificing friend!"
The lover accordingly gave up his life: like the rose, he played away his head, laughing and rejoicing.[33]

Rumi inserted some gaps in his narrative. Such gaps are characteristic features of the new structure of the texts that were introduced at that time. They break the rational linearity that is usually expected by readers. The first gap is inserted between Book 1 and Book 2. In the prologue to Book 2, Rumi refers to the gap as "an interval":

> Book 2 has been delayed for a while: an interval was needed so that the blood might turn to milk. Blood does not become sweet milk until thy fortune gives birth to a new babe. Consider well these words.[34]

The "gaps" are needed so a reader's mind is forced to pause at certain moments to absorb the previously administered impacts. Because it is necessary to absorb and then digest the initial impulse before the reader is ready to be exposed to more refined impacts.

A second gap is inserted at the conclusion of Book 6. Namely, the last story of Book 6 is ... not finished. In the final episode, Rumi only indicates that the lover "will win his prize completely."

Scholars have assumed that there should be a last, i.e., Book 7, in which the last story is completed. Some scholars even launched a

---

[33] *Ibid*, Book V, 1242-1270.
[34] *Ibid*, Book II, 1-2.

hunt for the missing book and got quite preoccupied with the search for this presumably lost volume.

Indeed, Rumi wrote Book 7. However, he used the same approach as that shown to Ibn-Arabi in his vision. Namely, the last book is like the missing chapter in Ibn-Arabi's *Bezels of Wisdom*. It forms the invisible frame-story within which all other six books are enclosed. At the same time, such a design introduces the second gap which is inserted between Book 6 and the hidden Book 7.

The purpose of the second gap is the same as the first one. The last volume is *available* only to those who correctly digest the impacts that are contained in the first six books. Otherwise, just like *Alif* with respect to the letters of the alphabet, the last volume remains invisible. It is hidden. The veil may be removed by absorbing Rumi's teaching contained in the first six books. Rumi pointed out that, as long as the readers need more words and letters to perceive his message, Book 7 would remain invisible to them, "If there are a hundred spiritual books, yet they are but one chapter."

Just like in the *Arabian Nights*, all stories of *Mathnawi* are interwoven into a frame-story. It was in the same manner that the changes within the Macrocosm were reflected in a new literary form. With this respect, Rumi's *Mathnawi* constitutes the second version of the *Arabian Nights*.

Mathnawi's design is based on a structure that is partially placed within the invisible or imaginary world. This structure is the same as that of the complex numbers. The complex numbers also include two parts, a real part and an imaginary one. Now we may recognize that it was such a structure that was shown to Cardano. And it was such a structure that served as the template needed to solve the cubic equations. This template was disclosed in Ibn-Arabi's and Rumi's writings in the 13th century. This means that it took some three hundred years before that template was absorbed, at least partially, in the human mind. Before that time, the concept of complex numbers could not be perceived by the rational mind.

Mathnawi influenced many generations of readers. It performed its function within its prescribed time. When the prescribed period expired, there was a need to introduce a more advanced literary form of Rumi's beloved. This newer beloved appeared in the 16th century. At that time a new beloved was needed to stimulate further the development of modern science.

First, however, the groundwork had to be properly prepared before a select group of physicists could be exposed to the effect of the *new* beloved.

The Craftsmen of Science

> To the Sage therefore Music and Color
> are but phases of the same expression.
> (*Fairfax L. Cartwright*)

As indicated earlier, the greatest physicists are either craftsmen or magicians. In physics, a craftsman approaches his task by following deterministic principles. He starts with a well-defined problem and a few initial clues. Then he follows them rigorously by applying a known mathematical theory. If he is successful, he ends up with a solution.

A physicist-magician, on the other hand, does not bother with intermediary steps. As we have seen earlier, he follows a clue given to him as a result of ... an unexplainable experience. In this way, he can come up straight with the solution. Afterward, he has to work backward to provide proof that his solution is correct. It seems, however, that the appearances of physicists-magicians are not incidental. Their appearances are correlated with changes occurring within the cosmic matrix.

Let us follow the trend that was initiated by the discovery of complex numbers and link it with the next milestone of the development of modern physics.

The foundation of modern physics was laid down at the time when Cardano wrote his *Book on Games of Chance* which was published around the year 1564. Coincidently, this first scientific treatment of probability was written in the year when Galileo Galilei, the father of modern physics, was born in Pisa, Italy.

Cardano's book on probability has on it our Genie's fingerprints. It seems that our Genie was not idle; he was already planning and working on his next trick. We may suspect that our Genie, in its subtler form, was behind the major events that led to the birth of one of the strangest scientific theories ever conceived by the human mind: quantum mechanics. Let us take a look at the overall circumstances that led to the formulation of modern physics. Just like in the case of the Italian mathematicians, the environment had to be correctly prepared before a seed could germinate. Such an environment included a certain time, specific places, and not-so-ordinary individuals.

The time was determined by the progress achieved by a materialistic science that followed strictly the rules of logic and was supported by data collected in ingeniously designed sets of experiments. This phase of the development was executed by several generations of dedicated craftsmen in classical physics. The entire effort was focused on the formulation of the laws governing the physical world and was driven by the desire to have a complete set of laws that would describe all matter from tiny photons and electrons to large objects such as planets and stars. This work required a careful collection of data and its organization to extract useful information. Organized information is what is commonly called knowledge and allows us to deal with data and facts that can be confirmed experimentally. At this point, it may be useful to differentiate between *knowledge* and *wisdom*. Knowledge does not affect the nature of man; it provides a set of data that allows us to understand the physical world. Wisdom, on the other hand, by its impact affects the very nature of ordinary man. It may bring man to a higher level of existence. By its very principle, science excludes such a possibility. Yet, it was *wisdom* that guided science at its most crucial moments.

Galileo studied gravity, the principle of relativity, astronomy, and worked in applied science and technology. He was the champion of the heliocentric system which brought him into conflict with the Roman Inquisition. In 1615, he was accused of heresy because his work was "foolish and absurd philosophy, and formally heretical since it explicitly contradicts in many places the sense of the Holy Scripture." Galileo published his analysis of several situations which demonstrated the principle of conservation of energy, including the movement of a pendulum. He used the pendulum to demonstrate how potential energy is converted into kinetic energy, and vice versa.

Galileo was followed by several generations of physicists and mathematicians. They laid down the framework for the formulation of deterministic science. Among those were Gaspard Gustav Coriolis, Thomas Young, James Clerk Maxwell, Heinrich Hertz, Max Planck, Albert Einstein, Sir Geoffrey Taylor, and Louis de Broglie.

In 1829, Gaspard Gustave Coriolis, a French mathematician, defined kinetic energy (E_k) in terms of the mass (m) and speed (v) of an object. In his textbook entitled *Du Calcul de l'Effet des Machines*, Coriolis came out with the kinetic energy equation:

$$E_k = \frac{m}{2} v^2$$

This equation indicates that the energy of a moving object is proportional to its mass and the square of its speed. Coriolis was an enthusiastic billiard player. He needed this formula for his mathematical theory of billiards, which he described in a book published in 1835.

Thomas Young was a British polymath who made notable contributions to the fields of light, physiology, language, musical harmony, and Egyptology. He has been called the founder of physiological optics. He developed the hypothesis that color perception depends on the presence in the retina of three kinds of nerve fibers. This foreshadowed the modern understanding of color vision, in particular, the finding that the eye does indeed have three color receptors that are sensitive to different wavelength ranges. As an Egyptologist, he contributed to the decipherment of the Rosetta Stone, a granite slab inscribed with a hieroglyphic decree issued in Memphis, Egypt in 196 BC. However, he is mostly remembered for an experiment performed in 1801 which led to significant implications for the future of science. The results of Young's experiment are relevant to the description of the modus operandi of the New Cosmos. It may be helpful, therefore, to get familiar with the principle of that experiment.

Young's experiment is known as a double-slit experiment. In this experiment, a beam of light illuminates a metallic plate with two parallel slits. The light passing through the slits is observed on a screen that is placed behind the plate. It was expected that the image on the screen would be in the shape of two spots, i.e., the optical projection of the two slits. Instead of the expected image of two slits, the light pattern on the screen was in the shape of a series of alternating bright and dark bands. Today, this shape is known as an interference pattern. It is a characteristic feature of waves. When waves emerge from two slits they interfere with each other. If their peaks coincide, they reinforce each other and they form a bright spot. If a peak and a trough coincide, they cancel out each other and they form a dark spot. Together, they produce a series of alternating bright and dark stripes on the back screen. The significance of Young's observation was that it indicated that light is a form of a wave. Starting with this experiment, the wave model was accepted for the description of light. This seemingly simple observation had a tremendous impact on the entirety of science and philosophy. At the time of its discovery, however, its significance was not fully comprehended. The time was not ripe yet.

James Clerk Maxwell, a Scottish physicist, was the first to bring together electricity, magnetism, and light and showed that they are three different manifestations of the same phenomenon. Maxwell's equations for electromagnetism have been considered to be the second great unification in physics after that described by Newton. He published his theory in 1865. The interesting thing was that Maxwell's equations contained the principle of the theory of special relativity, which later was formulated by Albert Einstein. However, at that time no scientist could recognize this feature yet. Again, the time was not ready for its full comprehension.

Two decades later, i.e., in 1887, the German physicist Heinrich Hertz added more mystery to the problem of the nature of light. He discovered an effect by which a beam of light was able to knock off electrons from metallic surfaces. This phenomenon became known as the photoelectric effect. However, the details of this effect were not consistent with the accepted theory of light. Namely, it was found out that ultraviolet rays were needed, while red rays did not display such ability. This behavior of light was contradicting the classical model of light. This was a sign that the properties of light were not fully understood yet.

In 1900, Max Planck presented the postulate that electromagnetic energy could be emitted only in equal and finite quantities called quanta. In other words, the energy E could only be a multiple of an elementary unit. This unit of energy was determined to be hf where h is the Planck constant and f is the frequency of the radiation. This postulate led to serious consequences. If a light source emits its radiation in quanta, then light should have a corpuscular structure. Yet, light, as demonstrated by Young, clearly propagates in the form of a wave.

This mystery was solved by Albert Einstein in 1905. At least it seemed that it was. Einstein explained the result of Hertz's experiment by proposing that light was not simply waves but discrete wave-packets of different frequencies. He showed that the energy of these wave-packets was proportional to their frequencies. He called them light quanta. Later, these *light quanta* became known as photons.

The excitement of solving the problem of light did not last very long. In 1909, Sir Geoffrey Taylor, an English physicist, repeated Young's experiment with a double-slit but ... with only one photon. The most striking result was when he placed a detector at one of the slits to find out whether a single photon was going through slit 1 or 2. In that case, the interference pattern disappeared! The photon appeared to behave like a particle. When more photons were sent, but one at a time, a pattern gradually emerged that looked exactly like that obtained in Young's experiment. This would indicate that a single photon behaves like a particle, but several photons are like a wave!

To make things even more confusing, electrons showed similar behavior. Electrons are tiny electrically charged particles. Yet, in the double-slit experiment, they also behaved like waves. The physicists were perplexed: what are we going to do with wave-like electrons?

Then, it got even worse. The next piece of the puzzle was added by Luis de Broglie, a French aristocrat. Born in 1892, Louis de Broglie was the 7th Duke de Broglie. De Broglie were French nobles. Their family name is taken from Broglie, a small town in Normandy. The family members were high-ranking soldiers, politicians, and diplomats. Luis de Broglie entered the Sorbonne in Paris. Following his family tradition, he intended to devote himself to the diplomatic service. When his father died in 1906, his elder brother Maurice, who was seventeen years his senior, became head of the family and responsible for Luis' education. They both lived in a mansion in Paris, where Maurice had a private laboratory for research on X-rays. When Maurice told him about the exciting developments in quantum theory and relativity, Louis gave up on his diplomatic carrier. Instead, he decided to study physics. In 1913, he received his B.Sc. degree and entered the army for his year of military service. When war broke out, he was assigned to the wireless telegraphy team that was stationed at the Eiffel Tower. He remained there for five and a half years, learning an enormous amount of practical electromagnetism. In 1920, he resumed his study of theoretical physics. He was attracted by the mystery of the structure of matter and of radiation that was introduced by the work of Planck. In his doctoral thesis, de Broglie extended the concept of particle-wave duality to material particles, especially to electrons. De Broglie's thesis was encapsulated in a simple equation:

$$\lambda = \frac{h}{mv}$$

De Broglie's equation implies that particles, e.g., electrons, may be looked at as waves, whose wavelengths (λ) may be calculated by knowing the Planck constant (h), their mass (m), and speed (v). This means that any moving particle may be looked at as a wave with a precisely determined wavelength. The conclusion of de Broglie's thesis was so unusual that the examiners were not sure about its validity. They sent de Broglie's thesis to Albert Einstein. Einstein enthusiastically supported the new concept and de Broglie's received his Ph.D. in Physics from the Sorbonne in 1924.

The mystery of light was staring straight into the eyes of the confused physicists. They were faced with a strange question: what or who was deciding that light sometimes was a wave and sometimes a particle? At this point in the history of science, the physicists were divided into two camps. The first camp claimed that matter was made from atoms and particles. The others insisted that matter was in a form of waves. Yet, these two areas of physics could not remain detached from each other. There was an urgency to unite them by the formulation of a theory of exchange of energy between particles and waves.

It was this question that led to the development of quantum mechanics. The discovery of quantum mechanics in the mid-1920s was the most profound revolution in physical theory since the birth of modern physics in the 17th century.

However, before we enter the fascinating world of quantum mechanics, let us pay a brief visit to the Paris Academy. It was there that, unknown and unrecognized by 19th century audiences, the solution to the mystery of quantum mechanics had been on display since as early as ... 1808.

A Travelling Scientist

> The world owes to him gratitude, since he made the sound visible.
>
> (*Wolfgang Goethe*)

The next episode of our story took place in Paris at the beginning of the 19th century. Although seemingly not related, this episode marks quite an important moment in our tale.

At the turn of the 18th and 19th century, some of the most attractive shows performed in many European cities were demonstrations of .... a scientific experiment. These demonstrations drew large crowds who were impressed by their sophistication and entertaining qualities. These shows were performed at academic institutions or in royal palaces in Dresden, Berlin, Göttingen, Bremen, Hamburg, Munich, Copenhagen, Flensburg; Danzig, Königsberg, Breslau, Riga, Petersburg, Tallinn, Prague, Vienna, Karslbad, Amsterdam, Brussels, and Paris. Let us attend one of these demonstrations.

It is December 1808. We are in the College des Quatre-Nations in Paris. It is there that the Paris Academy is located. It was created in 1666. It became a model for many other European academies, including those in St. Petersburg and Berlin.

The main hall of the College has been rearranged so that it does not resemble an auditorium. It looks rather like a large concert hall. Neither has the audience any resemblance with the usually somber academicians. There is a joyful and relaxed atmosphere. Women and men are elegantly dressed. All seats are already taken. Among the audience are the leading French scientists, including the famous mathematician Pierre-Simon Laplace and the equally well-known chemist Joseph Gay-Lussac. In the center of the hall, there is a podium with two strange-looking instruments. One of the instruments resembles a small piano. The other one is a square metal plate with a stand attached to its center. Beside the metal plate is a violin bow. The bow and the metal plate are placed on a small table. The chairman of the Academy introduces his guest as a "traveling scientist" who is considered to be the father of acoustics, a German musician and physicist, named Erns Chladni.

Chladni stands up and takes place next to the small table. He explains that he will demonstrate how to transform sounds into visible patterns. He sprinkles the metal plate with some sand in such a way that its entire surface is evenly covered. He takes the violin bow with his right hand and starts bowing along the edge of the plate. While bowing, he places two fingers of his left hand on the edge of the plate. The audience sees a beautiful symmetric pattern appearing on the surface of the plate. Then, he holds the plate with his fingers at a different place and the pattern is transformed into an even more sophisticated one. In this way, he can generate a series of highly symmetric and beautiful designs. The audience is fascinated with what they see. Some members of the audience stand up and applaud.

Chladni explains that the patterns are formed by the sand grains placing themselves along lines that do not vibrate. These lines are called nodal lines. He says that he has found the relationship between the quality of a sound ("pitch") and the various shapes of these nodal patterns. He has found that there are two basic patterns, linear nodes, and circular nodes. In this way, he was able to correlate each pitch to two corresponding numbers. This has allowed him to derive a simple formula to calculate the frequency of a sound by knowing these two numbers. He used this knowledge to design a new type of musical instrument. He points to the other instrument.

"This instrument, which I call a clavi-cylinder, is based on the principle that I have discovered with the vibrating plate. The sound produced by this instrument is generated by vibrating iron rods touching a set of rotating glass cylinders. The iron rods are activated by the keys and the cylinder is rotated by using a pedal. As you will hear, the sound is similar to that of an organ, because the duration of a note can be made to last as long as desired. However, this instrument is superior to the organ because the volume of each individual note can be increased or decreased by varying the pressure applied to the keys."

Chladni takes a seat at the clavi-cylinder and starts to play. He plays a composition that was specially written for his instrument by Joseph Haydn, an Austrian composer. The audience is mesmerized by Chladni's concert. They have never heard this sort of musical composition. It sounds like the mysterious music of heavenly spheres.

The audience gives Chladni another standing ovation.

Chladni identified forty-seven patterns which he presented in his book entitled *Discoveries in the Theory of Sound*. The "shapes" of these patterns are dependent on the boundary conditions of a medium within which sound waves are activated. In this case, the boundary conditions are determined by the geometry of the metallic plate. Wolfgang Goethe, whom Chladni met in Weimar, quite precisely summarized what Chladni did. Namely, Goethe wrote that Chladni made "sounds visible."

At that time, no one fully realized how significant Chladni's experiment was. The patterns formed by the grains of sand were a graphical illustration of various solutions to –what would become known as– the wave equation.

Hundred years later, Chladni's "table of sounds" was reproduced by physicists in the periodic table of chemical elements. The linear and circular nodes were substituted by the columns and rows of the periodic table. The arrangement of the chemical elements in the periodic table helped to find the model of an atom. Sometime later on, it was realized that Chladni's patterns were a visual representation of the process by which matter appears either as a particle or a wave. In other words, Chladni's "table of sounds" was another example of a solution waiting for a problem that would appear sometimes later in the future.

Yet, there was still much more that was revealed in Chladni's presentation. Chladni's experiment was a demonstration of the process of … the creation of the physical universe. It was a simplified but quite accurate illustration of how an invisible field is transformed into visible forms. Let us recall the description of the mechanism which, in accordance with the model of cosmic consciousness, led to the creation of the physical world:

> As the cosmic field oscillates, it creates series of waves within the boundaries of space-time. The waves are reflected back and forth from those boundaries. They form a set of standing waves, i.e., waves whose amplitudes are stable and do not move. The locations at which the amplitude of the standing wave has its minimum are called nodes, and the locations where the amplitude has its maximum are called antinodes. When the frequency of the oscillations increases, more sophisticated patterns occur. As a result, the entire field of consciousness is divided into regions bounded along nodal

lines. These patterns are molds for various forms of matter. The shapes and properties of various forms of matter are determined by corresponding templates within the higher zones of the cosmic field.[35]

In Chladni's demonstration, space-time is represented by the metallic plate. The bowing is equivalent to the oscillations of the cosmic field. However, the most significant feature of that experiment was the demonstration of the role of the observer. Namely, by changing the position of his fingers, the observer was able to transform one set of patterns into another set. Unknowingly, Chladni demonstrated the role of the "observer" in the manipulations of matter.

The question of the role of the observer in an experiment did not arise for another hundred years. This question, however, turned out to be the most troubling problem of modern science and led to a crisis in modern physics. Somehow, this aspect of Chladni's experiment is still beyond the perception of the brightest physicists. But ... we are getting a bit ahead of our story.

---

[35] *A Journey through Cosmic Consciousness*, W. Jamroz, p. 55 (see Note #1).

The Metaphysicists of Copenhagen

A synthesis embracing both rational understanding and the mystical experience of unity is the mythos, spoken or unspoken, of our present day and age.
(*Wolfgang Pauli*)

At the beginning of the 20th century, physicists believed that physics was the science of everything. They were convinced that if they could learn the laws that govern the elementary particles, the complete theory of the universe would be in their hands. They would be able to predict and explain everything that has happened and will happen in the physical world. Somehow, it did not occur to them that this type of knowledge cannot be carried beyond a certain point. No wonder that, at some point, the deterministic belief had to be challenged. At such a time it was necessary to inject a few fresh ideas which would force the physicists to re-think their approach and expand their conceptual horizons. In other words, it was time for the re-appearance of our Genie. Indeed, our Genie's intervention was a piece of art: the birth of quantum mechanics. With the discovery of quantum mechanics, all these rather naïve deterministic assumptions started to melt like a snowman on the first day of spring.

First, however, let us examine the overall understanding of the universe that prevailed among physicists at that time.

In the early 1900s, physicists were trying to come up with a formula describing the properties of photons and electrons with the same degree of prediction as that given by Newton's formula for large-scale objects. The experiment, however, demonstrated that electrons and photons did not behave at all in accordance with the classical model. Instead, they seemed to manifest a strange and incomprehensible characteristic. It was found that in the double-slit experiment, an electron behaved differently depending on whether it was observed or not. When observed, the electron behaved like a particle and passed through one or the other slit. When it was not observed, the electron behaved like a wave, passing through both slits. The physicists were mesmerized!

It seemed like an invisible genie was playing tricks with their measurements.

The physicists realized that they needed to change their approach entirely. A new type of equation was needed to handle the recently obtained data. An equation that would allow them to determine simultaneously the location of a particle and the shape of its wave. Afterward, they would have to figure out how these two are related and what is the factor that triggers the switch from the waveform into the particle form. It was then that a new group of actors arrived on the scene.

Not many people realize that it was a spark of mysticism that led to the discovery of quantum mechanics. This does not mean the discoverers were mystics or that quantum physics is part of mysticism. It simply means that the discoverers' minds were open to holistic concepts and ideas that were outside of the box of materialistic determinism. The discoverers of quantum mechanics were not afraid to expand their vision and allow other possibilities. It was that openness that led to the formulation of an entirely new view of the physical world. Some would even say that it was a revolutionary movement within the scientific community. However, after grasping the new concepts, these revolutionaries stayed solidly on the grounds of science; they applied the rules of deterministic physics.

This new child of science was born in the 1920s. It just happened that, at that time, there was a group of physicists who tried to explain the recent experimental data collected in laboratories in Denmark, Switzerland, and Germany. The main players were Niels Bohr from the University of Copenhagen, Max Born and Werner Heisenberg from the University of Göttingen, Wolfgang Pauli from the University of Hamburg, and Erwin Schrödinger from the University of Zürich. In other words, quantum mechanics was born in a predominantly German context.

Niels Bohr was a Danish physicist who, while still in his twenties, was the first to apply quantum theory to atoms. In July, September, and November of 1913 – he published in the *Philosophical Magazine* three papers which later became famous as "Bohr's trilogy." In these papers, he outlined the quantum model of the atom. Bohr, therefore, is considered to be the father of quantum mechanics.

Bohr's proposal led to an entirely new philosophy of science according to which neither particles nor waves are attributes of nature. They are just ideas formed in our minds. They are useful representations that provide intuitive pictures of invisible entities. They are imaginary extensions of known and observable objects which are familiar to us. Therefore, we impose forms, such as particles and waves, because these particular shapes are known to us.

Much of the debate about the new "philosophy" of science went on at the University of Copenhagen. Bohr called the new philosophy complementarity. The principle of complementarity holds that objects have certain pairs of complementary properties which cannot all be observed or measured simultaneously. This feature was quite foreign to classical mechanics. Bohr's position was quite radical. It was considered to be extremely anti-deterministic because he denied that it was even possible to describe an electron at all!

Bohr adopted the idea of complementarity from the Kabbalah, the Jewish mystical writings. The Kabbalah speaks of the complementarity between the Infinite and the finite universe. Bohr applied this concept to the relationship between reality and measurement, respectively. Later on, Bohr's theory became known as the Copenhagen interpretation. It says that a quantum particle does not exist in one state or another, but in all of its possible states at once. A particular state becomes reality only when it is measured. In other words, nothing exists until it is measured. It is through the measurement that potentiality is transformed into its manifested materialistic reality.

Bohr's concept attracted a group of physicists and mathematicians who championed his ideas. Bohr was their educator and mentor. In fact, he became their leader. Among his "disciples" were Werner Heisenberg, Wolfgang Pauli, and Erwin Schrödinger.

Werner Heisenberg was a German physicist. He was nicknamed "the Buddha" for his interest in India. In June of 1922, he attended the so-called Bohr Festival in Göttingen. At that event, Bohr was a guest lecturer and gave a series of lectures on quantum mechanics. It was then and there that Heisenberg met Bohr for the first time. The meeting had a significant effect on him. Later on, Heisenberg began an appointment as a university lecturer and assistant to Bohr in Copenhagen. Heisenberg was the first one to realize that a measurement has rather a dramatic impact on an electron. During a measurement, an electron is bombarded by highly energetic photons. As a result, the electron's state is changed. This means that there is a certain limitation inherent in nature that is imposed on the accuracy of measurements. Heisenberg expressed these limitations in an equation. The equation is known as the uncertainty principle. The Heisenberg uncertainty principle states that there is a fundamental limit to the accuracy with which both the position and the velocity of a particle can be measured.

Wolfgang Pauli was born in Vienna. Like Heisenberg, he met Bohr in Göttingen. For Pauli, this meeting was the beginning of a new scientific life. In the following years, he was appointed a lecturer at the University of Hamburg. Pauli was heavily influenced by Eastern mysticism. He was looking at mysticism as a synthesis between rationality and religion. He speculated that quantum theory could combine the psychological, philosophical, scientific, and mystical approaches to consciousness. Pauli also believed that the human mind could overcome the limitations of space and time. He was not only interested in breaking the limitations of time and space. He, as an individual, had a bizarre effect on his environment. He became known among physicists as a man able to break the most elaborate experimental set-up simply by ... walking into a laboratory. This was so evident, that his colleagues would avoid running their experiments if they knew that Pauli was somewhere around. There is a very well-recorded incident that occurred in a physics laboratory at the University of Göttingen. For no apparent reason, a sophisticated apparatus stopped working. Obviously, the head of the experimental group concluded that they had fallen victim to Pauli's effect. However, Pauli was not there. On this very day, he was on a train traveling from Hamburg to Zurich. However, later on, it was found out that, at that very moment, Pauli had indeed been in Göttingen: he was waiting for a connection at the train station there. Pauli was aware of his reputation and was delighted whenever this effect was manifested. These strange occurrences were in line with his collaboration with C.G. Jung on the concept of synchronicity. They both, Pauli and Jung, were carrying investigations into the validity of parapsychology. Afterward, Pauli's effect spread within the physicists' community and infected their laboratories all over the world. It is often used to "measure" the greatness of theoretical physicists. The greater a theoretical physicist, the stronger is his or her destructive effect on experimental set-ups.

The third member of the Metaphysicists of Copenhagen was Erwin Schrödinger. He was born in Austria. In 1921 he moved to Zürich. He had strong interests in Eastern religions, particularly in the Vedanta philosophy of Hinduism. He was fascinated with the possibility that individual consciousness is only a manifestation of a field that pervades the entire universe. Consequently, he believed that scientific work was an approach to the Absolute.

It was this group of physicists that gave birth to quantum mechanics. It all started during the Christmas holidays of 1925.

Tridents, Imp, Gunpowder, and a Cat

> Those who are not shocked when they first come across quantum theory cannot possibly have understood it.
> (*Niels Bohr*)

It is December 23rd, 1925. Erwin, a man in his early forties walks through a snowy street in Arosa. Arosa is a mountainous village in the Swiss Alps, located not far from Davos – a fashionable ski resort. Erwin is a professor of theoretical physics at the University of Zürich. The previous month, he had stumbled on a mathematical problem that had occupied him ever since. It was triggered by a seminar given by a bright mathematician, Peter Debye from ETH, the famous Swiss Institute of Technology. Debye gave an enthusiastic presentation of a new hypothesis on matter waves suggested by Luis de Broglie. Debye suggested that there should be an equation that would be able to encompass both characteristics of matter, i.e., corpuscular and wave-like. It was this suggestion that triggered Erwin's desire to derive such an equation.

The Christmas holidays gave Erwin the opportunity to escape from Zürich under the pretext of needing a break from the University and the city. His wife, Anny, decided to stay in Zürich. Erwin went to Arosa, his favorite holiday place. From Arosa, he wrote to his old girlfriend from Vienna and invited her to join him. While waiting for her arrival, he continued to work on his mathematical problem.

Anny, Erwin's wife, did not mind at all that her husband left her in Zürich. She had made her own plans for Christmas. She was to spend the holidays with Hermann Weyl, her husband's best friend. Hermann was the chair of mathematics at ETH. Neither did Hella, Herman's wife, mind her husband's affair with Anny. Hella was in love with Paul Scherrer, the head of the Department of Physics at ETH. Hella, a mercurial woman, was running a fashionable social salon frequented by scientists, philosophers, and their spouses. Hella's gatherings were well-known within Zürich's intellectual circles. Her parties served not only as a forum for heated intellectual discussions and exchanges of philosophical views. The guests of Hella's salon formed probably the first community in the modern world who openly enjoyed a sexual revolution and where extramarital affairs were not only accepted; they were commonly practiced.

On that day, Erwin decided to skip skiing. Instead, he stayed in his room at the Villa Herwig and worked. This was the first time in his entire life that he found something more exciting than skiing, hiking, or partying.

Near lunchtime, Erwin decided to walk to a nearby chateau where he could get a cup of excellent coffee with a freshly baked Viennese apple strudel – pastry named after the city of his birth. The café was located on the main floor. It was nearly empty, except for one table that was occupied by an elegantly dressed woman. When he walked into the café, the woman turned her face toward him and smiled somehow mysteriously. Erwin thought that he had seen this woman before, although he could not remember when and where. He had the strong impression that she had been waiting there for him. As he was passing by her table, she said:

"Professor Schrödinger, if I am not mistaken" – she addressed him rather coldly.

"Yes," he answered a bit surprised. "Have we met before?" he asked.

She made a gesture with her hand indicating that he should take a seat at her table.

"Did we meet at Hella's? – he asked her, taking a seat next to her.

"Did we meet at Hella's?" – she repeated his question with a smirk smile.

"I know we did!" – he said, feeling a bit irritated.…

As soon as he said that he somehow knew that it was the wrong thing to say.

"So, how needless was it then to ask the question!" – she shot back.

"It looks like you want to tell me something, don't you?" He tried to recover his balance.

"Only, if you will listen to me with patience and without interrupting me."

He felt like a schoolboy being tutored by a teacher.

There was something unusual about this lady; some aura of importance, or even haughtiness. But a natural one, not superficial or pretentious like that carried around by some of the aristocratic members of Hella's circles.

"All right, go ahead," – he gave up on trying to gain the upper hand in this strange encounter.

She became very serious and started to talk. She spoke with a slight French accent. He had an impression that she was delivering an important message:

"I will grant you a wish, just like Queen Mab in that latest children's story," she said.

"But think a bit before you state what your wish is. There can be only one wish, the wish of your entire life. So, tell me – what is your deepest desire?"

"What a joke," he thought. He looked at her carefully. There was definitively something special about her. Somehow, he felt a bit intimidated by her presence. "Then, let's play that stupid game," he thought, "and see what this is all about."

"This is very simple," he said. "What I really desire is to solve an incredibly sophisticated mathematical problem – and it would be beyond your comprehension for me to try to explain it to you," he said with a satisfying smile.

The expression on her face changed. She looked at him with sadness and disappointment.

"All right then. Although my own wish would be that your wish was a different one. But this is beyond the point." After a short moment of silence, she continued,

"Your wish is to find a solution to a nagging question suggested by Peter …."

"How do you know that?" he interrupted without even trying to hide his surprise.

"The deal is that you do not interrupt me, isn't it?"

"OK, I will keep quiet."

"You cannot find the complete answer. The time is not ripe yet; the human mind is not ready for it. But you may find a partial solution that will satisfy you and others, at least for a hundred years or so. In order to do this, you will have to distance yourself from what you know and from what you think that you know. You must try to free your mind from such thoughts; you have to go beyond them. At least for a few brief moments.

In a couple of days or so, you will arrive at a state when you will see clearly that 'what has to be preserved' will have to be augmented. You will be able to augment it by inserting Neptune's trident on both sides.

Afterward, you will have to transform the trident using Cardano's imp. And this will make you a hero!

You will then enjoy a period of glory and admiration. Sometimes later, when you will be basking in your fame, a colleague of yours will refuse to play a game of dice. Soon afterward, you will receive a letter with some gunpowder in it. You will feel like sending back a cat in a box. If you do so – then you will realize how little you know."

She stood up, took out of her purse a small pyramidal crystal, and put it on the table in front of Erwin. "Keep it – but be careful. There is a *giant* inside it."

She waved her hand saying goodbye. And she was gone. "Tridents, imps, gunpowder, and cats! What kind of nonsense is this! What in the hell does it mean!" He thought that the lady was just out of her mind. And he was angry at himself for getting into this stupid conversation. He was a bit irritated and ... a bit disappointed because, despite of all her weirdness, he found her to be incredibly attractive.

He walked back to his chalet. As soon as he sat in front of his notebook, he forgot about the incident. He got completely taken by his task: how to find an equation that would apply both to a wave and a particle. And it would apply in such a way that it would comply with the principle of conservation of energy. He did not make much progress that day.

The following day he again decided to skip skiing. His girlfriend did not show up. He worked all day without much success.

That night he woke up with the strange feeling that he had to start working right away. He opened his notebook on a page with Coriolis's kinetic energy equation. At this moment he heard the voice of the mysterious lady: "What has to be preserved will have to be augmented. You will be able to extend it by adding Neptune's trident on both sides."

Of course, it was the energy that had to be preserved. He wrote the equation for the conservation of energy, where the total energy *E* of a particle is equal to the potential energy *U* plus the kinetic energy as defined by Coriolis[36]:

$$E = U + \frac{m}{2} v^2$$

---

[36] In our tale, it is not necessary to grasp fully the meanings of these equations. It is sufficient to look at them as drawings or abstract illustrations and follow the way they are changing.

This was obvious. But how to augment it? "By inserting Neptune's trident on both sides"? Suddenly it occurred to him that the Greek letter Ψ was often compared to a trident because of its shape. Coincidently, it was this very letter that was used as a symbol of the classical wave function. He mechanically added the letter Ψ on both sides of the equation of the conservation of energy:

$$E\,\Psi = (U + \frac{m}{2} v^2)\,\Psi$$

This did not make any sense. There was no mathematical, physical, rational, or logical justification for such a derivation. No physicist would do such a thing: this was like comparing apples with oranges. By doing that, the wave function Ψ was debased of its classical meaning: it took on an entirely new and unknown denotation. He looked at it and tried to make sense of what was in front of him. Then he realized that he could use the second-order derivative of Ψ and de Broglie's equation to eliminate the velocity.[37] This must have been the moment when Erwin was able to free his mind from "what he knew." And it was during that moment that Erwin obtained the equation describing the energy of standing particle waves:

$$E\Psi = \frac{-\hbar^2}{2m} \frac{d^2\Psi}{dx^2} + U\Psi$$

---

[37] The derivative is a measure of change with respect to a specific variable. An example of a first-order derivative is velocity as velocity is a measure of the change of distance with respect to time. Acceleration is an example of a second-order derivative for it is a measure of the change of the first-order derivative (the velocity) with respect to time. In the case of Schrödinger's equation, the second-order derivative ($d^2\Psi/dx^2$) is the change of the function Ψ with respect to its location x.
In the classical description, $d^2\Psi/dx^2 = -(2\pi/\lambda)^2\,\Psi$.
The $\hbar$ ("h-bar") is the Planck constant h divided by 2π.

The history of science recorded this event as Schrödinger's "brilliant guess as to what the proper equation should be." Yes, it was a good guess!

This equation is one of the most important achievements of physics. It combines the properties of a wave and a particle. In the Copenhagen interpretation of quantum mechanics, the wave equation is the most complete description that can be given of a physical system. Physicists believe that this equation applies to the whole universe.

Now, Schrödinger had to work backward to provide proof that his equation would give correct results. He applied it to the energy states of an electron in a hydrogen atom. The solutions were well known as they were equivalents of Chladni's patterns. Schrödinger submitted the first draft of a paper entitled "Quantization as an Eigenvalue Problem" on January 27th, 1926, i.e., less than three weeks after his return from Arosa. In the paper, he showed that his equation gives the correct quantization of the energy levels of the hydrogen atom.

Until today, physicists and historians have not been able to find out how Schrödinger derived his famous equation. He left plenty of personal notes about many things, including autobiographical comments, his letters to friends, and even a chronological account of his pathway to his discovery. Yet, his own account of the derivation of the most famous equation of modern science "seems to be almost deliberately cryptic."[38] Somehow, even Schrödinger's personal diary for the year 1925 has also mysteriously disappeared. Some more intuitive researchers have suspected that the presence of a mysterious lady, like the Dark Lady of Shakespeare's Sonnets,

---

[38] *Schrödinger: Life and Thought*, Walter Moore (Cambridge University Press, New York, 2015).

might have to do something with it. Particularly that, according to Hermann Weyl, Schrödinger did his greatest work during times of romantic outbursts. However, efforts to establish the identity of this woman have so far been unsuccessful. Therefore, it has been concluded that the mysterious lady of Arosa will remain a mystery. But, let us go back to Erwin.

In his first paper, Erwin had treated the wave function of a stationary system. In its first form, his equation was applicable to the wave-like processes occurring within the boundaries of atoms. Now, he needed to add a time dependence.
This led him to complex mathematical expressions that involved fourth- and even sixth-order differential equations. He struggled with this for several months.
During that time, he realized that he lost the crystal that the mysterious Lady had given him. Somehow, the crystal disappeared. He thought that the crystal disappeared while he was still in Arosa.
It was in June that he recalled the mysterious Lady's comment about Cardano. He did not know what Cardano's imp was, but he knew that Cardano was known for the use of imaginary numbers for solving cubic equations. Erwin was facing a similar problem: how to reduce the higher-order differential equations. He decided therefore to use imaginary numbers.
He arbitrarily introduced a complex form of the wave function. This was a "trick" that made it simpler to find the solution. Up to this point, Erwin had been convinced that Ψ must be a real function. Now, however, he was troubled by the appearance of imaginary numbers in his theory. But he decided that the imaginary numbers could be avoided by simply taking only the real part of complex terms. The imaginary numbers, just like in Cardano's case, would be merely a convenient device for calculations. In the same ways as they were used in

the analyses of sound waves and electrical circuits. When he introduced imaginary numbers to lower the order of the differential equations, a miracle happened.

He published his results in June of the same year. Now, the entire quantum mechanics was in his hands. He started to bask in glory and admiration. Occasionally he felt a bit uneasy about it, for he remembered the words of the mysterious Lady, "then you will realize how little you know."

We may recognize that Schrödinger's equation was obtained in a similar manner to Cardano's solution. It was the typical work of a scientist-magician. The function Ψ could not be derived – it could only be conceived intuitively.

Schrödinger's case, however, was more sophisticated. In the case of Cardano, there was a magic trick with "i" – the unit of imaginary numbers. It helped to solve the problem and then it disappeared. In Schrödinger's case, it was the function Ψ. This function, however, insisted on remaining there.

The latest Schrödinger's equation provided a way to calculate the wave function of a system and how it changes in time. However, Schrödinger's equation did not directly say what, exactly, the quantum wave function Ψ was. And that was quite extraordinary. This was a serious problem, not just for him, but for all those working in the field of quantum mechanics. Here is an equation that is accepted as one of the greatest achievements of modern science, yet physicists still do not understand one of its key parameters! As of today, there is no one satisfactory interpretation of what the quantum wave function Ψ really means. How is it possible to be satisfied with a theory in which the major element is ... a mystery.

Initially, Erwin thought that the function Ψ was somehow related to an electric charge density. At that time, he did not yet fully appreciate that the function Ψ contains the hidden information that was essential for a complete description of the quantum world.

Erwin did not follow completely the Lady's hint about Cardano's method. Otherwise, he would have done the same with Ψ that Cardano did with "i": square it!

If Erwin did not do it, then somebody else had to do it.

Just a few days after Erwin submitted his paper with his latest version of the equation, Max Born submitted a report in which he did the "squaring." In this way, he was able to explain the meaning of the "square" of the quantum wave function. (This, however, did not explain the meaning of the Ψ function itself.) Now Erwin knew that he had to wait for one of his colleagues to "refuse a game of dice." He did not have a clue what that meant. But by now, he realized that the mysterious Lady of Arosa knew very well what she was talking about. Definitively she knew much more than any physicists at that time.

Max Born determined that the function Ψ was related to the probability for an electron to be localized in a specific region. More precisely, it is the square of the wave function modulus ($|\Psi|^2$) which gives that probability. The squared wave function indicates the probability that the electron is in or near a certain area. According to Born's interpretation, a quantum system remains a superposition of a multiplicity of probable states until it interacts with something or until it is measured. When this happens, the superposition "collapses" and turns into one of the possible states. It was at this point that determinism in physics came to a full stop.

The very nature of quantum mechanics may be illustrated by a game of dice. A dice may be used as a representation of a quantum system, such as an electron, photon, or atom. A traditional dice is a cube with each of its six faces marked with a different number of dots, from one to six. Let us assume that the dots represent the specific states of a quantum system. Therefore, there are six states which are marked as one dot, two dots, and so on – up to six dots, respectively. In other words, the dice is in all six possible states at the same time. Each of these states may be ascribed to a certain probability of realization. However, before the dice hits the table, none of the dice's states does really exist. Yet, all the possible probabilities are precisely described by the ($|\Psi|^2$) of the dice. Any of these states may become reality when the thrown dice hits the table. This moment corresponds to the measurement. At that moment, the wave function of the dice "collapses" and becomes known: only one of the six states becomes reality. Prior to that event, this particular state did not exist; it was only "probable." There was only a certain probability that this particular state could be manifested.

Whenever a measurement is made, the probabilistic state of a system is transformed into the rational world: the totally non-deterministic state is transformed into a deterministic one. Both descriptions, non-deterministic and deterministic, are needed in this marvelous relationship discovered by modern physicists.

Many physicists refused to accept this kind of interpretation of reality. For them, such an approach was too mystical and not precise enough. One of the strongest opponents of the statistical interpretation of quantum mechanics was Albert Einstein. He accused Niels Bohr of scientific heresy for introducing "mystic" elements in his interpretation of quantum mechanics. In one of his letters, Einstein famously wrote:

Quantum mechanics is certainly imposing. But an inner voice tells me that it is not yet the real thing. The theory says a lot but does not really bring us any closer to the secret of the "old one." I, at any rate, am convinced that *He* is not playing at dice.[39]

To which Niels Bohr replied:

Einstein, stop telling God what to do with his dice.[40]

When Erwin heard about Einstein's comment, he knew that it was Einstein whom the Lady of Arosa was referring to in her remark about a "colleague who was going to refuse a game of dice." As predicted by the Lady, Erwin was basking now in fame and admiration. However, he was not quite comfortable at all with the overall situation. Now, he himself did not fully subscribe to the Copenhagen interpretation. He felt more comfortable with Einstein's point of view.

---

[39] *The Born-Einstein Letters: The Correspondence Between Albert Einstein and Max and Hedwig Born, 1916–1955, with Commentaries by Max Born*, translation by Irene Born (Walker and Co., New York, 1971, p. 88).
[40] http://www.gmilburn.ca/2009/06/15/the-mystics-and-realists-of-quantum-physics/#:~:text=Winner%20of%20the%20Nobel%20Prize,view%20had%20no%20place%20in.

***

Einstein disagreed with the concept of probabilities as a description of reality. He strongly believed that reality was knowable and measurable. He felt that the quantum theory had abdicated the historical role of science to provide knowledge on aspects of nature that are independent of observers or their observations. Einstein tried to discredit the Copenhagen interpretation. In 1935, together with his two colleagues Boris Podolsky and Nathan Rosen, Einstein published a paper in which he proposed a thought experiment with the intention to bury the probabilistic interpretation of quantum mechanics and put it forever into its grave. It was a very clever argument that became known as the EPR experiment (from the first letters of its authors' names).

The crux of the EPR experiment was based on a system that emits two photons simultaneously. According to the quantum theory, the quantum state of each photon of the pair cannot be described independently of the state of the other. Both photons are described by the same wave function. In accordance with the angular momentum conservation law, at any point in time, the photons must have opposite spins. This means that if one photon has a spin up, the other photon must have a spin down. In the EPR experiment, these two photons are separated by a long distance. Now, if the spin of one photon is measured, the spin of the other becomes known instantly. In other words, the result of the measurement is *instantaneously* transmitted to the other photon. This was Einstein's famous "spooky action at a distance." (Later on, this effect became known as "quantum entanglement.") Quantum mechanics implied that this type of "information" would have to be transmitted faster than the speed of light. And this was in contradiction to the accepted Einstein's theory of special relativity. In the conclusion to their paper, the authors wrote: "We are thus forced to conclude that the quantum-mechanical description of physical reality given by wave functions is not complete." This was a winning argument; and, as it seemed at that time, it could not be disproven. Einstein appeared to be right: in nature, there is no room for a game of chance.

> Erwin was getting nervous. Now he was expecting ... a shipment of gunpowder.

A letter with gunpowder arrived a few weeks after the publication of the EPR paper. The letter was sent to Erwin by Einstein. In this letter, Einstein was proposing another thought experiment to demonstrate the inconsistency of the Copenhagen interpretation. This time, Einstein was describing an experiment with gunpowder that can spontaneously combust. The system may be considered as a quantum system that is simultaneously into two states, i.e., "not-yet exploded" and "already exploded." And that is obvious nonsense – because anyone can tell whether the system has exploded or not. Therefore, the entire business with the probabilistic nature of the wave function is simply ridiculous.

Now Erwin knew that it was time for him to "send back a cat in a box." At that time, Erwin was becoming more and more skeptical about the Copenhagen interpretation of quantum mechanics. Erwin did not need much time to work out his response. He wrote a paper that became widely known and often quoted as "Schrödinger's cat." It was another thought experiment to demonstrate how ridiculous the probabilistic interpretation of quantum mechanics was. In Erwin's thought experiment, a cat is penned up in a box together with a Geiger counter and a tiny bit of radioactive substance. The radioactive substance is so small, that in the course of an hour there is an equal probability that one or none of the atoms may decay. If an atom decays, the counter detects it and through a relay releases a hammer that shatters a small flask of hydrocyanic acid. The released acid kills the cat.

Before the box is opened, the wave function of the entire

system presents a quantum superposition of the cat that remains simultaneously "alive" and "dead." Such a possibility is, of course, impossible. It is obvious that the cat is either dead or alive, there is no other possibility.

In this way, Erwin intended to illustrate the absurdity of the Copenhagen interpretation of quantum mechanics.

Einstein and de Broglie happily welcomed Schrödinger as a member of the "realists' camp." The "realists" were fighting against the "mystical dogma" that strongly interfered with their deterministic beliefs.

Although the "realistic" view started to dominate the world of physics, there still was that uncomfortable function Ψ. It seemed that the function Ψ did not have any intention of disappearing. Physicists did not understand it. They did not realize that the Ψ function contained hidden information, which they had not been capable of recognizing.

Sometime later, Paul Dirac commented that the function Ψ contained an element that was well hidden in nature; it was so well hidden that, before Schrödinger's magical trick, people were not even aware of its presence. In other words, there was a new element that was introduced into the mind of early 20th century man, an element that played an important role in expanding human cognition.

Spectrum of Matter

> This confluence of the higher and lower worlds is part of the matrix that sustains life within current modes of reality. Access to these perceptive realms has often been activated through the power of creative imagination. Within perennial literature there has always been a distinction between what is 'true imagination' and what is fantastical imagination.
> (*Kingsley Dennis*)

Physicists have determined that, as far as the size of physical objects is concerned, there is a lower limit to the physical world. According to quantum mechanics, there is a certain length that determines the smallest possible size – beyond which the physical world cannot have any structure. The size limitation is derived from the fact that to measure a smaller structure, it would require so much energy within such a tiny volume that it would transform it into a black hole. Therefore, no physical matter can exist beyond that point. This lowest possible dimension is known as the Planck length. The Planck length is defined to be about 10^{-35} meter. This is an incredibly small value. Its order of magnitude may be visualized by considering the length of a football field (10^2 meters) and the size of an electron (10^{-16} meter). The Planck length is as much smaller than an electron – as an electron is smaller in comparison to the size of a football field.

The Planck length defines the lower limit of the quantum world. Therefore, quantum mechanics applies to the region above that physical limit. Physicists, however, have not considered the possibility of an upper limit of the quantum world. Instead, they *assumed* that quantum mechanics is applicable across the entire physical world. Such a view is an example of thinking within the paradigm of Zeno's paradox. And it is this kind of linear thinking that prevents us from understanding the basic principle of quantum mechanics.

This "linearity" was imposed by Einstein's gunpowder and Schrödinger's cat thought-experiments. In these thought-experiments it was assumed that the same type of wave function describes quantum particles and macroscopic objects. This assumption implies that there is one wave function that is applicable to elementary particles, an experimental set-up, and an observer. This view is consistent with Bohr's correspondence principle. Bohr's principle states that all physical objects must comply with the rules of quantum mechanics. The principle says that it does not matter whether we are dealing with electrons, tennis balls, or galaxies – their behavior must follow the laws of quantum mechanics. Consequently, quantum mechanics should be expandable and applicable across the entire universe.

Physicists accept that there is one parameter of quantum mechanics that applies to all objects regardless of their size. This parameter is the wavelength of an object. This wavelength is determined by de Broglie's equation, i.e., it is inversely proportional to the mass of that object. This means the heavier the object, the shorter its wavelength. In accordance with this view, we do not see quantum properties among macroscopic objects because the wavelengths associated with large objects are extremely small. The interference patterns for such wavelengths would be so tiny that they are simply not measurable. This is why, physicists claim, it is difficult to observe quantum effects among macroscopic objects.

***

Let us take a closer look at the quantum world from the perspective of the model of cosmic consciousness.

The quantum world is placed at the border with the invisible world. According to the model of cosmic consciousness, this border separates the quantum word from the transition zone where the oscillating field of consciousness is being converted into matter. It is there that elementary particles of matter are being "conceived." And it was in that zone that the Big Bang occurred. In other words, it is the zone where matter flows into space-time.

After passing through the transition zone, matter appears in its first physical form as the *quantum vacuum*. As indicated earlier in this book, the quantum vacuum is the lowest level of the physical world. In physics, it is referred to as "nothingness." The quantum vacuum contains bubbling fields within which various shapes of particles are popping in and out. It is there that elementary particles are "born" and form the quantum world.

In the quantum world, the matter is not quite stable; it appears simultaneously as particles and waves. This "instability" or "duality" are remnants of the oscillating nature of the field of consciousness. There are specific laws that rule over the matter within this region. These laws became known as the quantum theory. As particles start to form larger and larger structures, matter moves into the region where the gravitational force becomes predominant. As a result, the matter becomes more stable. At one point, matter loses its particle-wave duality. It becomes a particle-like substance. The rules that govern stable structures have been discovered as the laws of classical physics.

As the structures become more and more massive, the flow of matter enters the region where it forms huge objects. At one point, the objects become so large that they are not capable of sustaining their own inner structures. Once again, the matter becomes less stable. In this region, the matter collapses and gradually turns into a wave-like substance. This process has been described by physicists as the formation of black holes. The region where black holes are formed is next to the zone where matter is converted back into consciousness (C). This means that black holes are formed at the border with the invisible world. This region may be looked at as an equivalent to that of the quantum world; it may be called the macro-quantum world. The laws that govern it have not been discovered yet. The following graph illustrates symbolically the spectrum of matter:

In summary, there are three regions of matter.

1. The field of consciousness is converted into the matter in the region just below the Planck length. It is there that matter flows into space-time and forms the micro-quantum world. (At the present time, "new" matter has been observed by cosmologists as a flow of particles called "cosmic rays.")
2. Afterward, the quantum world is transformed into the macroscopic world, the world of classical physics.
3. At the higher end, matter forms the macro-quantum world. Subsequently, it is re-converted into the field of consciousness and exits space-time.

The entire spectrum of the matter is surrounded by the field of consciousness. There is a continuous flow in and out of matter coming from the field of consciousness. It is the lowest zone of cosmic consciousness that is in direct contact with matter. Physicists have detected the presence of this zone of consciousness and its effect on matter. They coined it "dark matter." (It is called "dark" because it does not interact with observable electromagnetic radiation; it is invisible.) Therefore, it may be said that matter is "floating" within dark matter. Dark matter provides an invisible framework within which the entire physical world is enclosed.

We may notice that the overall structure of the physical world is like a replica of the design of the *Arabian Nights*. The various regions of the spectrum of matter are like episodes in the life of the universe. The matter is conceived within dark matter. At its birth, it takes on the form of the quantum world. It reaches its maturity when it arrives in the classical world. In its old age, it ends up in a black hole. Afterward, it is converted back into dark matter. In this context, the black matter acts as the frame-story within which all the episodes are inserted. Without the frame-story, the episodes are incomplete. So it is with the physical world: without the field of consciousness – the scientific description of the physical world is incomplete. And just like in the second version of the *Arabian Nights*, the frame-story of the universe remains hidden from the rational mind.

It is this "hidden" aspect of the universe that has confused physicists. Despite discovering the lower limit, they have assumed that matter is governed by a set of identical laws across the entire physical world. In this way, they have ignored the possibility of discrete regions of matter. As there are three unique regions, therefore there are different sets of laws that operate within each of these regions. One must learn about the overall framework first before attempting to combine these regions into one coherent system.

***

The model of cosmic consciousness may also help to unravel the mystery of the wave function Ψ.

The quantum world is a specific region that is affected by its closeness to the border with the invisible world. This closeness to the invisible world is marked by the duality of matter, i.e., it appears simultaneously as waves and particles. These two different forms of matter, waves and particles, are coupled together. The ratio of this coupling is not fixed; it depends on its location along the spectrum of matter. Therefore, there is a need for a coupling function that would describe this quite extraordinary and sophisticated relationship. This coupling function has been identified as that "mysterious" wave function Ψ. The term "wave function" is a bit misleading because it implies that this function has wave-like physical characteristics. Its form is similar to the classical function. As pointed out earlier, when the quantum "wave function" appeared in Schrödinger's equation – it lost its classical meaning. Namely, a classical wave function describes the behavior of a particular physical field and has the dimension of that field. The function Ψ, on the other hand, is always non-dimensional. In its very nature, it is a real function. It contains information about the coupling ratio between wave-like and particle-like components. The coupling ratio is non-linear and, to make things more difficult, by its very nature it is probabilistic. And, as found by Max Born, the squared value of this coupling ratio ($|\Psi|^2$) gives the probability of finding matter in its particle-like form (or in its wave-like form) in a specific location at a given time.

The presence of the function Ψ is a unique feature of the region of matter that belongs to the quantum world. Outside of the quantum world, however, the function Ψ disappears. All objects which are outside of the quantum world lose their wave-like aspects. In other words, solid matter cannot be described by using wave parameters. This means that de Broglie's description of matter does not apply to classical objects.

***

It seems that the most intricate scientific discoveries are to serve, among other more obvious practical applications, as hints indicating that there is a much more "complex" science that governs the process of the development of the universe and the human mind. The human mind has the latent capacities to perceive the overall "frame-story" which is beyond the range of conventionally experienced physics. In this context, we may look at the laws discovered by modern physics as a reflection of the evolutionary potential that is locked in the human mind.

Who is the Observer?

Everything on Earth exists because it has its origin in another dimension, where that thing is perfect, where the multiplicity of forms is understood –not only perceived– as unity.
(Jalaluddin Rumi)

Einstein's, de Broglie's, and Schrödinger's views prevailed ... but only until 1964. In that year John Bell, an Irish physicist, published a simple and elegant equation that allowed to calculate a theoretical limit beyond which the observed effects can only be explained by quantum entanglement, not classical physics. The equation became known as the Bell Inequality. It was the Bell Inequality that validated the "spooky action at a distance." Later on, the Bell Inequality was proven experimentally. At that point, the scientific community had to accept that Einstein's and Schrödinger's thought-experiments were seriously flawed. Rather than a supposed mockery of quantum mechanics, the quantum entanglement turned out to be reality. The Copenhagen interpretation re-gained its status.

***

Today's physicists have given up on the possibility of resolving the mystery of some of the aspects of quantum mechanics. There is no problem with how to use quantum mechanics to calculate some of the most extraordinary features of matter. These mysterious elements did not stop engineers from exploring quantum mechanics. Their work has led to entirely new fields of technology such as quantum communications, teleporting devices, quantum cryptography, and quantum computers. Yet, the true nature of quantum mechanics is still perplexing and mystifying. Despite a century of thinking, physicists still do not really know how to deal with the fundamental aspects of quantum mechanics.

One of the most "mysterious" aspects of quantum mechanics is the role of the experimenter or the observer. This problem did not appear in classical physics. There was no place and no special status for the observer in Newtonian mechanics or Maxwell's electromagnetism. In the Copenhagen interpretation of quantum mechanics, however, there is an important role that is ascribed to the observer. According to the founders of quantum mechanics, the quantum state of a system is described by a quantum wave function that encompasses not only elementary particles but also includes the experimental apparatus and the human observer. This seems to indicate that the experimenter becomes part of the entire experimental set-up. And, as quantum mechanics is supposed to apply to the entire physical world, it is believed that such objects as planets and the entire universe may be considered as part of the experimental set-up. Such a belief invites various opinions which suggest the need for the inclusion of additional parameters such as, for example, human free-will or human consciousness. And it is at this point that quantum mechanics enters a controversy.

Most physicists agree that there is such a thing as consciousness – although they do not clearly define what that term means. Most of them are satisfied with vague and non-specific descriptions such as: consciousness is something that we all have; consciousness is part of nature; consciousness exists but quite apart from the laws of physics; and so on and so forth. Whatever consciousness is, they say, definitively there is no room for it in hard-core physics. Adding "consciousness" to the equation would greatly disturb the entire foundation of science. Science would lose its credibility.

It has been noticed that when there is an imprecision in the understanding of a concept, then simplified, naïve, and inaccurate explanations are popping out to fill the void. Not surprisingly, the vagueness of the role of the observer has led to simplified and misleading statements that provided fodder for New Agers and other groups in the enlightenment industry. It is quite common these days to find in scientific and pseudo-scientific literature such statements as:

- Nothing exists until it is measured.
- When we measure something, we are forcing an undetermined, undefined world to assume an experimental value.
- We are not measuring the world, we are creating it.

It was this sort of imprecise thinking that led to various fictitious ideas. It has even been suggested that physicists could create and re-create new realities by simply observing the outcomes of their experiments. These ideas prompted Einstein to ridicule this kind of interpretation by posing the question: "Does the Moon exist only when looked at it?" Again, Einstein was ridiculing such an obviously faulty interpretation. Yet, Einstein's question clearly showed his lack of understanding of the basis of quantum theory. To the last days of his life, he himself was not able to draw a conclusive picture of quantum mechanics.

Today's physicists have slightly modified their interpretation of the role of the observer. They accepted a *belief* that an observer is not needed to collapse a wave function. They say that there is something else in the universe that plays the role of the observer. The moment a measurement is performed, they say, the physical record of the new state of the particle is made by something, somewhere in the universe. According to this belief, the information is recorded even if no human was looking at it. Consequently, the wave function collapses because the physical universe has already experienced it. Therefore, the quantum measurement has nothing to do with the human observer, whether he or she is conscious or unconscious, a wise person or a fool. This belief has led to a widely accepted conclusion: physicists do not have to worry about the philosophical interpretation of quantum mechanics. They have happily abandoned it and left it up to others to deal with. With this sort of explanation, physics managed quite cleverly to get away from the conscious observer and human free-will. Yet, their interpretation is obviously incomplete. It does not specify what that "thing" is that makes and keeps a record of the measurement.

So, does it mean that it is not possible to come up with an explanation of this "mysterious" aspect that seems to infringe on one of the biggest intellectual achievements of mankind?

The main reason for this difficulty is not so much some complicated mathematical equations or some seemingly metaphysical effects. The main obstacle is the conceptual barrier in the mind of 21st century man. It is the same type of barrier that we saw in the case involving cubic equations or in the challenge of formulating the quantum wave equation. Just like in the past, this obstacle may be overcome as soon as the human mind accepts certain concepts which, as of today, seem to belong to sheer fantasy or overcharged imagination. This situation has been quite precisely summarized by Steven Weinberg, a contemporary physicist. He intuitively perceived that physicists are more like hounds than hawks: they have become good at sniffing around on the ground for traces of truth, but they do not seem to be able to see the overall pattern of the truth.

It is the limited approach of a sniffing hound that imposes limits on the mind. One would have to combine both modes of operation, that of a hawk and that of a hound, into one to get out of the present impasse. Or, if expressed in terms of modern psychology, it would mean to escape from the domination of the left hemisphere of the brain. It does not mean, as some psychologists tend to imply, delegating the leading role to the right hemisphere of the brain. Instead, it is required to bring both modes of operation, holistic and sequential, into a balanced and harmonious interplay of the two. In this way, it is possible to activate supersensory faculties which are superior to the rational mind – because they can reach towards the "complex" Cosmos. "Complex" means consisting of an entangled interplay between the physical world and the Macrocosm.

***

As found out by physicists, the lower limit of the quantum world is precisely determined. It is defined by the Planck length. What about the upper boundary of the quantum world? As we have seen in the previous chapter, at one point the quantum world is transformed into the classical world. Therefore, this prompts the question: is there a feature that could be used to determine the upper limit of the quantum world?

Yes, there is. There is a parameter that determines the region within which the quantum world is confined. This unique parameter of the quantum world is … quantum entanglement, the famous "spooky action at a distance." The quantum world is limited to an area within which the effect known as quantum entanglement may be activated and preserved. This means that at the edge of the physical world there is a region where it is possible to experience some extraordinary things. Particles that "live" in that region can acquire new properties. These properties allow these particles to be united with others in a special way. Namely, when they are united, they become "one." Then, they seem to know instantly about the states of their partners; they can share information with each other instantaneously. This effect can be observed only within the quantum world. When these particles become a part of larger structures, they lose this capacity. At this point, they cannot be entangled anymore. The structures of large assemblies of particles are too "rigid" to be affected by their closeness to the invisible world. It is then that the particles leave the world of probabilities and enter the classical world of deterministic physics. In other words, entanglement is a key factor that determines the limits of the quantum world.

The most interesting aspect of the quantum world is that, unlike its lower limit, the upper limit is not determined by a fixed length or size. This is because quantum entanglement is not a "given" property of quantum particles. This property is latent. Certain experimental conditions are needed to activate it and display it. This is a copy of the human mind. The human mind contains a set of latent supersensory faculties. When activated, these faculties allow one to expand the limit of ordinary human perception. In the same manner, the properties of an assembly of elementary particles can be augmented. In this context, we may view entangled particles are a replica of the latent faculties of the human mind.

The upper limit of the quantum world is determined, to a certain extent, by the might of human ingenuity. It is at this point that human consciousness is getting involved in quantum mechanics. The existence of an upper limit allows for the determination of the role of an "active observation," i.e., an act that induces the collapse of the function Ψ. Before we elaborate any further, let us first recall some basic features of the model of cosmic consciousness. As we have seen earlier, the model of cosmic consciousness provides an overall framework that helps us to explain these seemingly "mysterious" features of the quantum world.

There is a gradient of consciousness across the entire physical universe. Within the visible world, humans occupy the highest zone of consciousness, whereas the quantum world belongs to the lowest zone of that world. Between humans and the quantum world, there are several zones that are occupied by minerals, plants, animals. These various zones are placed at different positions along the gradient of consciousness; they are quite different qualitatively. Each of the zones operates through its specific wave functions. These wave functions are not interchangeable; they are specific to each zone. Therefore, the wave functions that describe the quantum world do not apply to the objects which belong to the zones of consciousness that encompass minerals, plants, animals, or humans. This is the reason why we cannot observe quantum effects for larger objects; this has nothing to do with de Broglie's wavelength being too small.

According to the model of cosmic consciousness, an act of active observation may be performed only on objects that belong to lower zones than that of the observer. There must be a certain minimum "gradient" of consciousness between the observer and the object of observation for the act of active observation to take place. It turns out that within the physical world there is only one possibility that meets this condition. Namely, such a minimum "gradient" occurs between the highest and the lowest zones of the field of consciousness that govern the physical world. The highest zone is occupied by humans; the quantum world belongs to the lowest zone. This means, that ordinary humans may act as active observers but only for objects that belong to the quantum world. And it is this condition that allows the rational mind to intervene in the quantum state of elementary particles.

Now, let us take a closer look at the mechanism of the act of observation. It turns out that it is not so much an idle observation that contributes to the collapse of the wave function. Instead, it is the experimental set-up that ultimately induces the collapse of the wave function. The experimental set-up and the implemented measurement methodology act as an overriding matrix for the region of the field of consciousness which governs the quantum world. It is the "matrix" of the experimental set-up that enforces elementary particles to behave in a certain way. It is through the experimental set-up that the collapse of the wave function is induced by the experimenter. In other words, the experimenter by designing the experimental set-up and implementing the measurement methodology can manipulate the quantum world. At the moment of the commencement of the experiment, the matrix of the experimental set-up becomes infused with that specific region of the field of consciousness. It is this region that is affected and, at the same time, "registers" and "keeps" a record of the measurement. This is the crux of the entire quantum process. Therefore, the "conscious observer" means the designer who conceives and activates the experimental matrix that collapses the wave function. As soon as the experimental matrix is activated, the experimenter does not have to be there to observe it. It is the experimental matrix that collapses the wave function.

We may recognize that the overall concept of the "observer" was nicely demonstrated by Erns Chladni. By changing the experimental matrix by repositioning the attachment points, Chladni was able to "collapse" the acoustic waves.

Recently published data indicate that quantum entanglement has been induced in materials consisting of thousands of atoms. However, such objects as gunpowder, Geiger meters, cats, and planets do not belong to the quantum world. If an object is outside of the boundaries of that specific region of the field of consciousness, the rational mind is incapable of manipulating it. This is why the Moon is there regardless of whether we are looking at it or not. It is this feature of the quantum world that escaped Einstein, Schrödinger, and others.

It is probably unstoppable that, sooner or later, human ingenuity will attempt to entangle more and more sophisticated systems. This will provide the groundwork for quite advanced computers, holographic communication systems, and other devices which will be capable of processing unimaginable amounts of data. It is quite realistic to think about the possibility of fabricating entangled automatons. However, all future attempts at entangling physical objects will be limited to that lowest region of the field of cosmic consciousness within which the quantum world is confined. Therefore, it may be said that entanglement determines the extent and the limit of intervention into nature by the rational mind. Richard Feynman, a distinguished contemporary physicist, intuitively perceived its importance. He is on the record as saying, "this effect has in it the heart of quantum physics; in fact, it contains the only mystery." Quantum entanglement is that qualitative factor that determines the limits of human free-will as far as physical matter is concerned. It is quite ironic that, by attempting to mock quantum mechanics, Einstein unintentionally introduced a concept which turned out to be one of the greatest secrets of nature!

Physicists have found out that they may partially manipulate the physical world. Within the quantum world, the rational mind may exercise its "creative" ability. However, there are strictly defined limits for such human interventions. The ordinary human mind's intervention is limited to that tiny region that is defined as the quantum world. Therefore, such a creative ability marks the rational mind's potential and its limits. It is important to emphasize the fact that not every human can perform such an action. Every human has the potential to do it, but to perform such a function, a person would have to go through specific learning and training. This means that such an ability is not "given." But it may be acquired. The existence of such possibilities is to show mankind the overall principle of creation and the role that man may play in it. The augmentation of the quantum world is a replica of the formation of the New Cosmos.

Like all major scientific discoveries, entanglement in its presently known form is a symbolic illustration of much more profound cosmic processes. It is quite astonishing to realize that the entire process of collapsing a wave function is a simplified but precise copy of the process of ... creation.

***

The existence of an upper limit to the quantum world will require a reformulation of Bohr's correspondence principle. Namely, the principle should consider an object's entangleability, i.e., ability to form entangled systems. In its redefined form, the correspondence principle would state that:

> Quantum mechanics is applicable to those objects which are mechanically entangleable.

The model of cosmic consciousness may be useful for drawing further conclusions from such a redefined principle. It allows for the clarification of those seemingly mysterious features of quantum mechanics. Firstly, quantum entanglement is a characteristic feature of a specific zone of the field of consciousness. As there are various zones of the field of consciousness, so there are qualitatively different processes leading to the formation of entanglements of the various zones along the gradient of cosmic consciousness. Consequently, it is possible to reach towards ability of entangling human minds through specific macrocosmic zones. In this case, the macrocosmic matrix is equivalent to the "experimental set-up." Therefore, it may be said that a particular macrocosmic zone together with the human minds which are entangled with it – forms an "entangled mind." Inducing such an entanglement is quite a sophisticated process that is outside of the competence of today's science. Yet, such a process is not only possible. It is necessary for the sustainment of humanity and the universe. As we have seen in the previous chapters of this book, it is through entanglement that there is a continuous exchange between the physical universe and the Macrocosm. Yet, such a possibility remains hidden for a majority of people. It is going to take some time before this potential of humanity is fully recognized.

Physicists' *belief* in the completely deterministic nature of physics served them very well through hundreds of years. Today's physicists, however, would have to abandon their classical and simplistic determinism in order to overcome the obstacle that they themselves have discovered. They would have to modify their belief and accept that certain things cannot be measured by their sophisticated apparatus. They would have to work with models based on parameters and relationships which, by definition, cannot be observed experimentally.

Like in the past, overcoming the current obstacle will require a sort of mental "adjustment." Such an adjustment does not appear randomly or accidentally. It is part of a very precise process. This is why, from time to time, at critical moments of the evolution of the human mind, there is a need for our Genie to appear and unveil a fragment of the bigger perspective. According to this book's overall assumption, some hints about entanglement had to be implanted and had to appear in literature prior to its recognition by physicists.

Are there any traces of its appearance in literature prior to Schrödinger's work?

Yes, they are.

The Western Lover

> In each mirror, each moment the Beloved
> shows a different face, a different shape.
> (*Fakhruddin Iraqi*)

With the discovery of complex numbers, our Genie was getting closer and closer to his freedom. At the same time, a more subtle entity was needed in literature to transmit the impulses that had been made available within the Macrocosm. In Rumi's writings, such a role was invested in the "beloved." As we have seen earlier, Rumi's "beloved" was not an ordinary person. She was a symbolic representation of an evolutionary impulse that, when absorbed, would activate the faculties of supersensory perception. Since then, "love" became a prevalent theme in mystical poetry.

Not everyone had a chance to experience an encounter with such a "beloved." Even if one did, there was no warranty for a satisfactory outcome. The process was becoming more and more demanding and sophisticated. So was the entire ambiance that surrounded such encounters. The presence of more advanced macrocosmic impulses was manifested by new kinds of effects and events which could not be explained within the boundaries of psychology, philosophy, or science. Obviously, such effects would be looked at as a sort of fairy tale or products of naïve imagination. It is quite reasonable to assume that, for example, any literary illustration of a "spooky action at a distance" would be considered as belonging to this category.

As the focus of the evolutionary process was being shifted from Central and Western Asia to Western Europe, there was a need to inject the set of new concepts and ideas that were projected from the Macrocosm at that time. Based on the previous episodes of our tale, it is obvious that these sorts of changes do not appear randomly. There is always a preparatory phase that is initiated several centuries earlier. And, usually, the new concepts and ideas are first disclosed in their symbolic form in literary sources.

In other words, a Western version of the *Arabian Nights* was needed.

This means that Rumi's "beloved" had to go through an adaptation. The updated "beloved" would not be tailored in accordance with the eastern patterns that dominated Rumi's writings. Neither could she be a copy of the classic Greek and Roman heroines. The new "beloved" would be specially tailored for the future Western mind. Her appearance had to be adjusted accordingly. She had to be fully compatible with the cultural and intellectual environment of Western Europe.

As the next chapter of physics was going to be written in Europe, the new version of the *Arabian Nights* would have to be written in one of the European languages and set-up entirely within a European cultural framework. It seems that 16th century England provided the needed environment. The political situation in England offered another advantage. Namely, it was possible to express the latest phase of the evolutionary process entirely within a secular context. Consequently, the third version of the *Arabian Nights* was released at the end of the 16th century in England. It was there that the original literary prototype of the Lady of Arosa appeared for the first time.

So, what was this new version of the *Arabian Nights*, and in what form was it delivered?

First, however, let us look at the overall state of western civilization. Western civilization was, and still is, heavily influenced by materials inherited from ancient Greece. This applies to philosophy, literature, and science. However, the form in which Greek philosophy reached the West was not quite complete. The materials adapted from the Greeks were not fully understood. It was assumed that Greek philosophy aimed at the development of intellectual skills. The emphasis was on rhetoric, logic, and various forms of intellectual exercises. In reality, however, this particular portion of Greek philosophy aimed to indicate … the limitations of the intellect, to caricature indulgence in sterile thinking. Zeno's paradox quoted earlier in this book is a good example of such a caricaturized approach. Zeno's paradox was used to emphasize the limitations of logic and linear thinking. To fulfill its overall objective, the speculative or rhetorical part of the philosophy would have to be fused with the practical methods of the development of the mind. Otherwise, it would be meaningless as far as any true progress of the development of the human mind is concerned. In the original Greek teaching, there was another portion that was focussed on the practice of self-development. That portion was left out. Without its practical portion, philosophy is just idle intellectual gymnastics, a caricature of true development of the mind. Just as it is spelled out by Hamlet in his warning:

> There are more things in heaven and earth, Horatio,
> Than are dreamt of in your philosophy.
> ("Hamlet" I.5)

Without its practical aspect, philosophy could not help man to free himself from transient things which otherwise paralyze and manipulate his mind. In other words, the form of philosophy that was adopted from the Greeks – was developmentally sterile.

The same applies to the literary forms that the Europeans inherited from the ancient Greeks and Latins. For example, Greek mythology, as it is known now, is incomplete. It is a collection of tales but without the background against which it was originally composed. It lacks its original overall frame-story. The Greek myths adopted in the West may be compared to randomly selected tales from the *Arabian Nights* but without the main frame-story. The key reference point is missing.

The third version of the *Arabian Nights* was released to fulfill these gaps.

***

The third version of the *Arabian Nights* was released at the end of the 16th century. It was written by William Shakespeare and was presented as 37 plays that were released over a time period of twenty years or so. As the new version reflected the most recent changes within the Macrocosm, its design was much more sophisticated than the previous collections.

Following Rumi's design of *Mathnawi*, the overall frame-story of the third *Arabian Nights* was hidden within those 37 plays; it was seemingly invisible. The frame-story was structured in accordance with the most recent design of the cosmic matrix. The other challenge for readers was the fact that the plays were released at random; the order in which they were made originally available did not follow the sequence of the overall narrative. It was left up to the readers to recognize both, the sequence of the tales and the hidden frame-story. To recognize the overall frame-story, the readers would have to overcome quite a number of conditionings that were implanted in their minds. Let's remember that the main purpose of the third *Arabian Nights* was to prepare the human mind for its exposure to the new concepts and ideas that would allow, among other things, for the development of modern science. The human mind had to be prepared for such an adjustment. The first step was to recognize the mental conditionings that acted as barriers preventing its expansion.

The overall frame-story of the third *Arabian Nights* is an illustration of the history of Western civilization. However, this is not history as it is known from school classes or university courses. The history that we learn from scholars is a collection of arbitrarily selected episodes that are arranged in accordance with the preferences of a particular historian. The history described in the third *Arabian Nights* is an account of the development of the human mind as seen from the perspective of the Macrocosm. In other words, this is a "complex" history that describes how the macrocosmic structure was reflected in events taking place within the ordinary physical world. It is a chronicle of the evolutionary process; a record of human struggles as mankind was trying to keep up in step with the changes occurring within the Macrocosm. This history is concerned with the overall purpose of humanity, while ordinary history is driven by man's inferior preoccupations.

***

The main character of the third *Arabian Nights* is a "beloved." It is the quality of the "beloved" that serves as a marker of the evolutionary growth of Western civilization. This "beloved" is more sophisticated than her predecessor described by Rumi. She takes on different appearances during the various stages of the evolutionary process. The new beloved's appearance depends on the maturity and quality of the environment within which she appears.

The first tale (play) of the third *Arabian Nights* is set during the Trojan wars. This is the starting point for all other episodes. It provides the background of the circumstances that existed prior to the initiation of the process that led to the foundation of Western civilization. Symbolically, it represents the state of the mind of the ancient world. At that time, the human mind was in a state of chaos. This was the result of events that occurred in antiquity. At one point in the remote past, there was a temporary break in the projection of the macrocosmic matrix. In its nature, this break was similar to that which occurred prior to the episode with Noah in *Genesis*. As a result, the evolutionary process was disturbed. This led to the appearance of corrupted faculties of the mind. These corrupted faculties were represented by men and women with seemingly god-like powers who were entirely driven by their untamed egoistic and sensual tendencies. These men and women abused the extraordinary responsibility that they had been charged with. Instead of overseeing the evolutionary process, they focused their activities on pursuing inferior objectives. They started to act and behave as demigods and demigoddesses. In other words, their appearance marked the corruption of the process. It was their actions and doings that were recorded in Greek and Roman mythologies and are symbolic illustrations of the consequences of the evolutionary failure. The contamination led to the Greek Dark Ages.

It is such a degenerated situation that is illustrated in the first episode of the third *Arabian Nights*. The corruption of the process is emphasized by the absence of a "beloved." Instead, heroines who appear in the first episode are presented as "whores" and their partners as "cuckolds." This was the state of the human mind at the time when a new phase of the evolutionary process was initiated to remedy the situation.

Because of its corruption, ancient Greece could not be used as an adequate vehicle for the initiation of the next stage of the process. A new infrastructure had to be built. In accordance with the cosmic matrix, the new phase of the evolutionary process was to be implanted in Western Europe. Rome was selected as the receptacle for the projection of a renewed evolutionary impulse. In order to discharge this function, Rome had to be linked to the macrocosmic matrix. Secondly, it would have to assimilate the renewed evolutionary impulse. There are four episodes that illustrate Rome's struggle with its evolutionary task. The Roman "beloved" is presented as a subdued and silent lady. At first, she is ignored. Then she is abandoned and even hurt. At the end, she is taken out of Rome and transferred into another milieu. The third *Arabian Nights* indicates that Rome failed to assimilate the available evolutionary energy. The process was abruptly brought to a stop. Just like ancient Greece, Rome was unable to fulfill its evolutionary potential. In ordinary history, the Roman fiasco was marked as the European Dark Ages.

Following Rome's failure, an alternative approach was needed to continue the process. Ancient Celtic Britain was chosen as an alternative milieu. The Celtic branch served as a backup for Rome. Accordingly, the previously released evolutionary impulse was transferred from Rome to 1st century Britain. There are three tales that illustrate the process implemented within the Celtic branch. The "beloved" who appears in the Celtic episodes is much stronger than her Roman predecessor. She can take actions on her own; she is trustful and incredibly loyal. However, her designated partner is still not quite ready for her. He is boastful, naïve, and distrustful. It is at this point that he is sent for his "training" into the future. From 1st century Britain he is sent to 13th century France and then to 16th century Italy. While traveling in the future he is exposed to much more subtle evolutionary impacts which were supposed to soften up his still raw nature. In both places, he meets the "beloveds" who were present there at those future times. As a result of these encounters, he is changed. However, the change was not sufficient to comply fully with the evolutionary needs of the Celtic branch. When he is back in 1st century Britain, he is still incapable of appreciating fully the quality of his "beloved." Once again, the process is interrupted. Consequently, the Celtic "beloved" is transferred to another place and to another time.

The question may be asked why such a sophisticated "journey" into the future was not successful. The answer to this question is related to the balance between free-will and causality. Namely, causality at the level of an ordinary man is the field of operation for the macrocosmic matrix. Any adjustment to the evolutionary plan is manifested on the level of the ordinary man as a series of intertwined opportunities. Each of these opportunities gives man a chance to make constructive choices. From the macrocosmic perspective, this is all that can be done. It is left up to man to make decisions. It seems that ordinary mankind is mostly driven by tendencies towards destructive choices. According to the third version of the *Arabian Nights*, it took some fourteen centuries before the impulse of evolutionary energy, which was invested in the 1st century Britain, could be assimilated within the ordinary world.

After the failures of the Roman and Celtic branches, the next evolutionary effort was once again focused on Greece. In the ordinary world, this effort was marked by the formation of Byzantium in the 4th century AD. Byzantium served as the next evolutionary option that was offered to mankind after the Roman and the Celtic failures. The Byzantine Empire was the continuation of the Roman Empire. It is also referred to as the Eastern Roman Empire. Byzantium continued to thrive for more than a thousand years. During most of its existence, this empire was the most powerful economic, cultural, and military force in Europe.

At this point, a tale is inserted that serves as a subframe-story for the first part of the third *Arabian Nights*. The events illustrated in this tale summarize the first phase of the process that led to the formation of modern European society. This subframe-story encompasses the episodes that took place in ancient Greece, ancient

Britain, Rome, Byzantium, and medieval France. These episodes are described from the point of view of a macrocosmic observer.

As mentioned earlier, the origin of evolutionary corruption was related to an event that took place in ancient times. Therefore, the correction of something already actualized in time could be accomplished only by an entity that was not only outside time but also outside existence. This implies that the corrective action had to be realized within the Macrocosm. Let us recall that all events taking place within the physical world have their origin within the macrocosmic matrix. The matrix contains the entire template of whatever did happen in the past, whatever happens in the present, and whatever will happen in the future. In accordance with the overall concept of the New Cosmos, the activities of humans may affect, to some degree, the structure of the macrocosmic matrix. Therefore, it is possible to change the matrix in such a way as to correct some of the errors of the past. In this manner, it is possible to reduce, at least partially, the effect of past failures.

The lover who appears in the subframe-story is a Prince. He is a "perfect lover." He can travel within the Macrocosm. He travels back into the past and then into the future. In this way, he can correct some errors that were made in the past. During his travels in the Macrocosm, he meets his "beloved" and then brings her with him onto the physical world. The new "beloved" brings with her the entire spectrum of evolutionary impulses that are available in the Macrocosm. After their arrival within the ordinary world, a new phase of the evolutionary process is initiated. Up to this moment, the third version of the *Arabian Nights* paralleled the second version. Starting from this point, the narrative diverges from its previous version. It illustrates the new phase of the process that becomes much more sophisticated than those implemented previously. Namely, the process is split into four evolutionary branches which are activated in medieval Europe. Each branch contains a partial evolutionary charge which was invested in Western Europe. These branches are described in four sets of parallel narratives which are included in the second part of the third *Arabian Nights*.

The new branches are activated in the Kingdom of Bohemia, France, England, and Italy.

The Bohemian branch took its roots in Byzantium. From Byzantium, the Bohemian branch continued westward. There are six tales that illustrate the process implemented there. The action of these tales takes place in the various regions of Central Europe, between the Adriatic Sea and the Baltic Sea. One of the tales is set in Bohemia. This is why these six episodes are called the Bohemian tales. At one point, there was a modification of the "beloved." Namely, the "beloved" was split into two impulses represented by twins, a young woman and her brother. Such a transformation was a sign of a significant change within the macrocosmic matrix. The presence of two impulses allows for the simultaneous activation of two new subtle faculties of the mind. In ordinary history, these events were manifested as the European Reformation that was initiated in Central Europe. The last episode of the Bohemian tales points out that the new form of "beloved" could not be sustained; the impulse was withdrawn. In the ordinary world, the withdrawal was manifested as the partial fiasco of the Reformation. However, exposure to this impulse allowed for the effective training of a future "lover."

The French branch was initiated in the early 12th century. There are three French tales in the collection. At that time, there was an attempt to accelerate the process. A certain risk was taken by skipping a few intermediate stages of the process. The French branch contained an evolutionary charge consisting of four impulses. The charge is represented by four beautiful young women. This means that there was a possibility for simultaneous activation of four subtle faculties of the human mind. This was an attempt to make up for the delays that were caused by the Greek, Roman, and

Celtic failures. Once more, however, the intended partners of the four "beloveds" were incapable of recognizing them. These men were still "blind" to the true quality and potential of the situation in which they found themselves. These men were still after trivial gains. The last of the French tales indicates that the attempt was only partially successful. It was not possible to fulfill the available potential within the French branch. However, it was possible to transfer the evolutionary charge to England.

The English operation was initiated at the same time as the French branch. There are eleven tales that describe the process that was implemented in England. From the very beginning, there were a lot of difficulties within this branch. Somehow, the overall environment of England was very resistant to this kind of adjustment. The most striking feature of that branch was the absence of "beloved." It was not possible to implant an evolutionary impulse there. However, there was continuous interaction between the English and the French branches. The situation was resolved in the last English episode. In the last episode, four evolutionary impulses were transferred from France to England. At the same time, one of the "lovers" was brought from the Bohemian branch. This allowed for noticeable progress. In the ordinary world, this evolutionary gain was registered as the birth of the English Renaissance.

The fourth branch was implemented in Italy. It sprouted from the Bohemian branch in the 13th century. There are seven tales that illustrate the various stages of the process in Italy. Within this branch, it was possible to activate two new faculties. This is illustrated as the marriages of two couples taking place at the same time. This union was marked in ordinary history as the birth of the Italian Renaissance.

The concluding tale of the third *Arabian Nights* summarizes the entire narrative. It is there that all four branches converge. All the branches are linked together. The final episode illustrates the evolutionary state of the Western civilization at the end of the 16th century. The union formed consists of four couples. This might be taken as a sign of success. In reality, however, this was only a partial success. The union of the four couples was formed but with the help of some artificial "patches." These "patches" were described symbolically as a sort of "magical potion." It looks like 16th century Western European man was still incapable of sustaining the available evolutionary charge. The lovers were still driven mostly by egoistic and sensual attractions. However, it was possible to patch-up temporarily the lovers' shortcomings. The patching-up was needed to give the European society a diluted taste of such an experience. A fuller experience of that particular union had to be delayed until some future time.

Since the time of their release, the tales included in the third *Arabian Nights* have served as part of a needed remedy for the Western mind. The collection was designed in such a way as to provide the evolutionary impact necessary to keep the European society on the course of its evolutionary growth. For the last four hundred years, this "impulse" was preserved. When read, these tales provide numerous impacts that prepare the mind for more effective assimilation of the available evolutionary charge.

***

So, what all of these have to do with our understanding of quantum mechanics? Quite a bit.

Let us take a look at an episode that is inserted into the third *Arabian Nights*. In one of the Bohemian tales, there is a most remarkable description of the effect that was discovered by physicists three hundred years later. This effect became known as "entanglement." However, that which is described in the tale is a much more sophisticated version of the effect. It is much more sophisticated because it illustrates the working of the entanglement of certain faculties of the human mind. As pointed out earlier, this kind of entanglement applies to the higher zones along the gradient of cosmic consciousness.

In this tale, two twin brothers illustrate the working of entanglement in the development of the human mind. The brothers' *twinness* symbolically indicates their entanglement. Just like in a "spooky action at the distance," after their birth, the brothers are separated and sent to two different locations. One of the brothers ends up on an island and grows up there. In this symbolic illustration, an island represents a higher state of the mind, i.e., a state that operates within the Macrocosm. This means that the "islander" symbolizes a subtle faculty in its latent form. The other brother remained within the physical world. He represents an ordinary faculty of the mind. When the islander reached the age of eighteen years, he felt compelled to be united with his lost brother. He started to look for him. Before he arrived in the place where his twin brother lives, he traveled on the sea for seven years. Let us recall that there is a different quality of time in the Macrocosm. Therefore, the "seven years" of the islander on the sea correspond to his brother's "seven days" in the physical world. During "these seven days" the earthly sibling went through some difficult and confusing experiences. In accordance with the entanglement principle, the ordinary faculty was affected by the experiences of its entangled partner. The entire tale illustrates how, through entanglement, an ordinary mind may benefit from experiences taking place within the Macrocosm. Entanglement is the basic principle that is being used in the development of the human mind.

***

At the end of the 16th century there was a subtle but significant breakthrough in the development of the human mind. The tales included in the third version of the *Arabian Nights* illustrate a structure of the mind that is much more subtle than that described in the available writings of Plato, Aristotle, Cicero, Seneca, or other classical writers. The third *Arabian Nights* was the first literary source that described this particular design. Therefore, it is impossible to grasp the meaning of these allegorical illustrations by analyzing them through the prism of classical writers whose intellectual and philosophical resolution was too coarse to handle the subtleties of these tales. This is why today's scholars have great difficulty coming up with a single and coherent interpretation of the tales. Instead, they focus their attention on biographical details of the man whose name appears as the author of the third version of the *Arabian Nights*.

The Dream

> ... We are such stuff
> As dreams are made on
> (*William Shakespeare*)

The Theory of Everything (TOE) is the physicists' Golden Fleece. This dream is the main driving force behind their efforts towards the formulation of, what they like to call, the ultimate triumph of human reason. This is underlined in the titles of some of their books, for example, Steven Weinberg's *Dreams of a Final Theory* or Steven Hawking's *The Dreams That Stuff Is Made Of*. It is assumed that this theory would allow them to "know the mind of God."

Many physicists believed that their ultimate objective would be realized by the end of the 19th century. This did not happen. Then, in 1928, after the discovery of the equation describing the electron, Max Born announced that "Physics, as we know it, will be over in six months." He meant the completion of the Theory of Everything. This was nearly one hundred years ago and, according to leading physicists, today we are not much closer to the fulfillment of that goal. Nevertheless, there is still a lot of optimism that such a theory is just about to be arrived at any day. Definitively, it is going to happen, they say, within the present generation of physicists.

At the present time, physicists have developed the so-called Grand Unified Theory that describes three out of the four known forces: the weak nuclear force, the strong nuclear force, and the electromagnetic force. These three forces appear within the quantum world. However, the gravitational force is not included in this theory.

The current theory of gravity remains a classical theory. Gravity still does not belong to the quantum world.

According to the accepted Big Bang theory, the formation of the universe was driven by quantum mechanics. Therefore, the birth of the universe took place within the quantum world. In other words, the matter appeared first within the quantum world. In accordance with Bohr's correspondence principle, gravity must be compliant with quantum mechanics. So far, however, physicists have not been able to demonstrate that gravity is describable by quantum mechanics.

As indicated earlier in this book, the non-compliance of gravity with quantum mechanics is consistent with the model of cosmic consciousness. Sooner or later, physicists will be forced to face the fact that there is a spectrum of matter. And, consequently, the quantum world and the classical world belong to different regions of that spectrum.

As of today, the most promising approach to TOE is a model based on the so-called string theory. At first, the string theory was derived in the 1960s as a possible model for the strong nuclear force. It turned out that this theory was unsuitable for that problem. Soon afterward, it was suggested that the string theory could be a promising candidate for a quantum theory of gravity. Most physicists saw the string theory as the latest hope for the Theory of Everything.

The overall idea of the string theory is based on Einstein's postulate that space-time is the medium that transmits gravity. In accordance with this postulate, space-time would be flat if there was no matter at all. It is the presence of matter that causes space-time to warp. Stars, planets, the Sun, the Earth, the Moon – contribute to the warping of space-time. And it is this warping that transmits gravity. In this model, the planets follow their orbits because they are stuck in channels that are made out by warped regions of space-time.

The overall approach of the string theory follows the paradigm of Zeno's paradox. It extends Einstein's postulate by adding tiny curly warps of space-time that are caused by the presence of microscopic particles, such as electrons, protons, photons, quarks. In other words, this theory assumes that the gravitational force of elementary particles influences the quantum world. The underlying feature of the theory is the presence of an additional layer of matter that is made of strings. It is assumed that the strings are much smaller than the tiniest known particles. They form the most fundamental layer of matter.

Physicists theorize that the size of a string might be just above the Planck length. Unlike a particle that occupies a point in space-time, a string occupies a one-dimensional line. The strings are formed in the shapes of either open or closed loops. Open strings may join with other strings and form a closed loop. Similarly, a string can be divided into two pieces. These two types of geometrical transformations are equivalent to absorption and emission, respectively.

The strings' corresponding space-time is a two-dimensional surface in the shape of a cylinder or a tube. This surface is called a world-sheet.

Strings can vibrate in different patterns. If a string vibrates slowly, it produces a single "tone." Each "tone" represents a specific elementary particle. If the vibration is increased, then two "tones" are generated. Therefore, by increasing the frequency of vibration more particles are being formed. In this way, the entire spectrum of elementary particles may be produced. The interesting feature of the strings is the fact that they keep vibrating continuously; they do not lose their vibrating energies. The string theory suggests that these vibrations produce all the richness of the physical world.

We may recognize that the overall concept of the string theory is an extended version of that discovered by Ernst Chladni. This concept was demonstrated in that famous experiment performed by Chladni in Paris in 1808. The various "tones" produced by the strings are a more sophisticated version of Chladni's patterns. Instead of patterns of sound, these different "tones" produce the entire gamut of elementary particles.

Although seemingly simple and quite attractive, the string theory needs quite a complex framework to accommodate all the possible theoretical configurations. Strings, like sounds, are affected by the geometry of the space within which they vibrate. Just as the sound vibrations are controlled by the shape of the musical instrument, so the vibrations of the strings require an adequate container. Theoretically, it is possible to accommodate the string model by using an imaginary space known as Hilbert space. Physicists have estimated that in addition to the four-dimensional space-time, they would need seven additional dimensions. Therefore, the string theory requires eleven dimensions! And this is a serious problem for physicists. It is a problem because it is impossible to design an experiment that could be carried out within such an imaginary space. Therefore, there is no means to verify the theory experimentally. The problem of multi-dimensionality remains the stumbling block for any future advancement of this theory. This limitation is another form of indeterminism that haunts physicists since the 1920s.

Let us try to pinpoint the critical feature of the situation. The string theory attempts to combine classical physics with quantum mechanics. This means that it intends to bridge the classical deterministic domain with the probabilistic quantum world. It aspires to bridge these two qualitatively different regions of the spectrum of matter. In other words, it attempts to enforce quantum entanglement on the macroscopic world. This, as we have seen, cannot be done.

We may easily recognize that the current impasse is just a repeat of those encountered previously with the cubic equations and the quantum wave function equation. Furthermore, we may identify a certain pattern within the sequence of these obstacles. Our Genie had to appear to suggest how to use imaginary numbers to solve the cubic equations. The Lady of Arosa was needed to provide a hint on how to derive the quantum wave equation. It is somehow obvious to assume that something or somebody will be needed to indicate how to resolve the current impasse of modern physics. As we have seen in all previous cases, the solution was always made available several centuries prior to its actual discovery. Of course, it was always described in its symbolic form and inserted into easily available literature. Then, it was waiting to be discovered and applied. Like the seed of a flower, it was waiting for the time when the overall conditions would be ready for it to germinate, sprout out and manifest itself. Therefore, it is reasonable to assume that the solution to the current impasse related to TOE is already available somewhere in literary sources.

So, where is it and what is it?

A Magic Carpet and Aleph

> There is an almost uncanny persistence and durability in the tale which cannot be accounted for in the present state of knowledge. Not only does it constantly appear in different incarnations which can be mapped ... but from time-to-time remarkable collections are assembled and enjoy a phenomenal vogue, after which they lapse and are reborn, perhaps in another culture, perhaps centuries later: to delight, attract, thrill, captivate yet another audience.
> (*Idries Shah*)

The third version of the *Arabian Nights* was presented within a secular context. Yet, it contained many references to the *Bible* and the classic Greek and Latin literature. Such a background was a "given" for a 16th century European audience. However, these references were used to underline some of the flaws and limitations of the commonly accepted interpretations of those sources and to point out their inadequacy for a relevant illustration of the process of the development of the human mind. In their essence, Shakespeare's plays illustrate the transition from the coarse and ordinary into a much finer substance. Shakespeare, however, neither explained nor disclosed the design on which his plays were set-up. He just used the design that was projected at that time. And, just as Rumi's writings were obscured by those who heavily relied on dogmatism and religiosity, so were Shakespeare's plays obscured by the scholar's incomplete grasp of ancient Greek and Latin sources.

The question may be asked: what about now? Is there any updated version of the *Arabian Nights* that is applicable to us? Are there any references that would help to comprehend the content of Rumi's and Shakespeare's writings in the context of the 21st century?

Yes, there are. There is a fourth version of the *Arabian Nights*. The fourth version has been adapted and adjusted in such a way as to be adequate to the contemporary audience. Instead of classics and philosophy, the current version is presented within the context of modern science and psychology. Coincidently, like the third version, the fourth version of the *Arabian Nights* has also been written in 37 "episodes." These episodes include the complete background and references needed to grasp the current state of evolution of the human mind. They are described in 37 books written by Idries Shah which were made available between 1956 and 2003. All of them were published originally by the Octagon Press, a publishing house founded by Shah. Shah described and explained the evolutionary process of the human mind. Of course, the projection of the macrocosmic design has been further advanced during the last four hundred years. Consequently, a further refinement of the structure of the tales was needed.

Let us take a look at a few examples of these latest adjustments. This will help us to comprehend their relevance to understanding the current stage of the development of the human mind. The first example is "The Tale of the Hunchback," a well-known tale that was included in the first version of the *Arabian Nights*. Here is one of the versions of the tale from that earlier collection:

> Once upon a time, a certain tailor went into his garden and found a dead body. It was the corpse of a young man called Shakl, who had been a most successful and respected citizen of his town. The tailor was distressed to see that there was a crowbar beside the corpse. It looked as if this apparently respectable man was in fact a burglar, who had died while exercising his trade, and that all his good works would now be lost.

The tailor said to himself:

"In order to enable Shakl's good name and the continuing effect of his repute to be prolonged, I shall place his body near his house. People who find it will think that he went out for an early morning walk and collapsed near his home."

So he started to drag the corpse towards Shakl's house. Hearing a noise, however, he turned back and placed the body in the garden of the house on the other side of his. It was the house of a merchant. The tailor knew that Shakl had sometimes visited the merchant, so he left the body there, thinking that people would assume that Shakl had died of natural causes during a visit there.

But the merchant, hearing a sound in his garden and fearing burglars, opened a window and threw a heavy club in the direction of the noise. When he went out, he found the body of the dead man. The merchant assumed, of course, that he had killed him.

So he dragged the body to a nearby house. It was the house of the steward of the royal kitchen. The merchant propped the corpse up against the wall of the steward's house.

Now, the steward's house was always infested by cats and mice who, he was convinced, stole his butter. This night when he returned home and lighted his candle, he was startled at the sight of a man leaning against the wall of his kitchen.

"Aha!" he cried at the sight of the supposed thief.

"To think that all this time I've blamed the cats and mice when it was you!" He knocked the thief down with a mallet. The man fell to the ground and the steward thought that he had killed him.

He quickly took up the body and carried it away through the deserted streets to the marketplace. There he leaned the corpse up against the wall of a shop and ran away.

Soon after, a king's courtier was passing by the market. Earlier in the evening, someone had stolen the courtier's turban. When he spied a man leaning against the wall, he thought the man was wearing his turban. Imagining that this was the thief of his turban, he knocked him down with a resounding blow. Then he started crying and calling out for the watchman of the marketplace. The watchman appeared and ran to stop what he thought was a fight between two men. When the watchman discovered the dead man, he dragged the courtier to the governor and accused him of murder. The next morning, the governor gave orders for the hanging of the courtier.

The gallows were set up in the central square of the city, and the executioner prepared to hang the courtier. But just as the rope was being tied around the courtier's neck, the steward pushed his way through the crowd, crying:

"Do not hang him. I killed the man," and he told how he stroke a deadly blow in his garden.

When the governor heard the steward's story, he ordered: "Hang the steward."

But at that moment, the merchant ran to the gallows and cried:

"Do not hang him! I killed that man." And he told his tale.

"Hang the merchant!" cried out the irritated governor.

But then the tailor ran to the gallows and shouted:

"No! I am to blame!" And he told his tale.

The governor was further confused and immediately ordered the executioner to hang the tailor.

In the meantime, the king was told about the case. The king sent his Chamberlain to stop the execution and bring all involved before him. When the king heard the whole tale, he ordered that it be inscribed on parchment in letters of gold.[41]

And here is an updated version of the same story. This version is included in Idries Shah's book entitled *Knowing How to Know*. Let us

---

[41] An unabridged version of "The Tale of the Hunchback" may be found in a collection entitled *Tell Me a Story*, adapted by Amy Friedman (e.g., https://products.kitsapsun.com/archive/).

start with the episode when the merchant finds Shakl's body in his garden:

> (...) So the merchant dragged the body to a nearby house. It was the house of the steward of the royal kitchen. The merchant propped the corpse up against the wall of the steward's house.
>
> As he was going to leave a thought struck him. He woke up the steward and said, "I have just seen this body in your courtyard, and it is evident that you have killed this important and good man, Shakl. Swear an oath of loyalty to me, and give me half of your money, or I'll bear witness against you!"
>
> The steward agreed, and the merchant took the body back to Shakl's house, where he left it at the gate. When Shakl's parents went out of the house, which they shared with him, they were pleased to find that the corpse had returned: because in fact Shakl had died the night before, in their presence, from an ordinary heart-attack. A burglar had stolen the body and tried to throw suspicion on the tailor, who was his enemy. The burglar had dropped the body and his crowbar in the tailor's garden.[42]

In his commentary on this story, Shah explains that *Shakl* means the "shape of a thing." It is a symbolic representation of the macrocosmic design. Shakl's parents are those who brought the design into operation. At that time, this particular form of the design became sterile; it had already fulfilled its purpose in that shape. It was due to be interred. So, this was marked by Shakl's death. The tailor is a symbol of the people who continue to be impressed by something which has in fact no function to perform. The merchant

---

[42] *Knowing How to Know*, Idries Shah (The Octagon Press, London, 1998, p. 272).

acted as he did because he could not face the truth, and because it often does not pay to do so. The steward represents those who can be promised and threatened by things that have no reality. Even the merchant was confused, because, while he was watching Shakl's house to see what happened when the body would be found, he could not understand why Shakl's parents, seeing the body, clapped their hands with joy. The only people who knew what had happened – were Shakl's parents. All the others –the tailor, the merchant, and the steward– continued for the rest of their lives to believe either what they imagined had happened or what the evidence of their own senses had told them.

As we may see, the previous version of the tale is not quite complete. It does not contain all the details. With this respect, the tale is sterile. Readers of this tale, therefore, are not exposed to the impact which could help them recognize such patterns in their everyday lives. Those who are not aware of these patterns will always believe that things are as they obviously seem to be.

Here is an example of how some of the original tales have been sterilized by changes to the storylines and removal of those episodes which, from the point of views of translators and publishers, were either not aesthetic enough, moralistically confusing, or too odd for their taste. Instead, some sentimental and more pleasing elements were added. This story is known as "Abu and the Caliph":

> The Caliph of Baghdad, Harun al-Rashid, often traveled about the city in disguise and it was on one of those tours that he met Abu, a poor man with a rich imagination who loved to brag; he called his bragging poetry.
> The Caliph decided he wanted to play a trick on this tale-teller. The next morning when Abu woke, he discovered he was lying in a bed in a palace, with servants surrounding his bed. "Protector of the People," one of the servants said, "what

would you like to eat?"

"I must be dreaming," Abu said, but other servants drew him a bath, and one of them asked what he would like them to do for him. "Anything you wish," they said.

"Punish the thief named Ali," he said. Ali had cheated him and all his neighbors, and Abu could not stop thinking about that.

"It shall be done," the servants said as they handed him fine clothes to wear.

By the time his servants led him to the throne, Abu was certain he was the Caliph.

"It must have slipped my mind," he thought, and he smiled, for being the Caliph felt grand.

When the Grand Vizier strode in, Abu felt somewhat alarmed. Would he notice that Abu was not the Caliph and toss the imposter out onto the street? But the Grand Vizier bowed before Abu and said, "Protector of the People, you were right. The man called Ali has been cheating everyone. But how did you know?"

For a moment Abu was silent, but he recovered and said, "Last night I dreamed I was once a poor man named Abu who wrote poetry, and Ali cheated me. Go to Abu's house and find his mother. Give her the money Ali stole from her."

"It will be done," said the Grand Vizier.

When Abu finished giving orders, he yawned, and the servants carried him to a fine bed covered in silk. A servant offered him a glass of wine. He could not resist, and within moments he collapsed into unconsciousness.

The real Caliph who was disguised as a servant – laughed heartily.

Harun Al-Rashid was amused by many things, but this man Abu amused him more than most. "I want him to live here," he said. "But first we must finish our trick." The servants wrapped Abu in a rug and carried him home.

When Abu woke the next morning, he called for his servants,

but an old woman appeared and asked, "What do you mean calling for servants?"

"Where am I?" Abu asked, staring at her. "Who are you?"

"You're home in your bed," she said, "and I am your sweet mother who raised you. Where were you yesterday? I needed you and you were gone!"

Abu laughed. "I am the Caliph of Baghdad. I belong in my palace."

He got up from his bed, but Abu's mother barred his way. "Enough nonsense," she said. "Enough of your wild imagination. How dare you leave me yesterday when the Grand Vizier came to visit and handed me a bag of gold. What was I to do or say?"

Abu clapped his hands. "I sent the Grand Vizier to repay you the money Ali stole! So you see, it is true. I am the Caliph of Baghdad!"

Abu became more and more confused. "First I was Abu, and then I was the Caliph, and now I seem to be no one at all." He fell asleep uncertain about who he could possibly be, and when he woke, he was amazed to see the Grand Vizier standing before him. Beside him stood another man wearing the robe and the slippers and the turban of the Caliph of Baghdad.

Abu rubbed his eyes. "You must be the Caliph of Baghdad," he said. "And if you are he, then that woman must be right. She must be my mother, and I must be Abu."

The Caliph of Baghdad laughed. He enjoyed playing tricks, but he liked this man better than tricks, and so he set him free and explained the whole story. And then he invited him to live in the palace.

From that day on, Abu lived in the palace, and he married the Grand Vizier's daughter. Never again was he sad and he always understood that even better than being the Caliph was being himself, Abu, the happy man with the marvelous imagination.[43]

Now let us read a version of the same story in which the original inner structure has been restored. This version is included in Idries Shah's collection entitled *Seeker After Truth*. In this version Abu appears as Hasan:

Once upon a time, in Old Bagdad, there lived a man named Hasan, who was for long contented with his lot. He lived a harmonious life, attending to his affairs and looking after his small shop, which his mother helped him to run.

But, as time passed, he became uncertain as to the drift and direction of his life. "Is there not more?" he asked himself, and wondered, especially when sitting in contemplation in the evenings, whether he might not experience more and achieve what was possible to him.

Because he gave voice to these thoughts, certain men of religion in his locality were pleased to brand him as a free-thinker and malcontent, saying "Dissatisfaction is another word for ingratitude, and aspiration is a veiled term for greed; surely Hasan should be denounced by all right-thinking men!"

People listened to the narrow-minded clerics, and were annoyed by Hasan's questionings, and presently he could find few who would bear his company for very long. Even those who would listen to his ideas were confused by them; and Hasan regarded them as shallow people in any case.

So Hasan took to wandering away from his shop and sitting, towards evening, at the crossroad at the end of his street, to

---

[43] A full version of "Abu and the Caliph" is available in a collection entitled *Tell Me a Story*, adapted by Amy Friedman (see Note #41).

ponder his desires and to think over the hostility of the supposed wise men.

One day it happened that the Caliph Haroun al-Rashid, Commander of the Faithful, was on his nightly rounds in disguise, accompanied by his faithful minister Ja'far and his black executioner, the eunuch Masrur, when they came upon the huddled figure at the crossroad.

"There must be something to us in this man," said the Caliph to his companions. To Hasan he said:

"May we spend some time talking with you, as we are travelers who have completed our work in this city, and are without friends?"

"Willingly," said Hasan, "and as you are strangers you shall come to my house, where I can entertain you better than at a crossroad."

The four made their way to Hasan's house, which he had equipped quite luxuriously for the entertainment of guests – though he hardly ever had any– and passed a pleasant evening.

"Friend Hasan" said the disguised Caliph, "now that we are so well acquainted, tell us something of your desires and of your likes and dislikes, to while away the time."

"Honourable Sir and kindly traveler" said Hasan, "I am really a rather simple man. But it is true that I would like one thing, and I dislike one thing. I would like to be Commander of the Faithful, the Caliph himself, and I dislike above all the contemptuous and small-minded self-styled men of religion who make it their business to harass those who are not as hypocritical as themselves."

When an opportunity presented itself, the Caliph slipped a dose of powerful narcotic into Hasan's drink. Within a few moments, he was unconscious and the powerful Masrur carried him back to the palace.

When he came to himself, Hasan found that he was dressed in imperial garments, lying on a silken couch, in the palace of the Caliph, with minions massaging his hands and feet.

"Where am I?" he cried.

"In your palace, O Commander of the Believers!" chorused the attendants – for this is what they had been ordered to do by Haroun himself.

At first, he could not believe that he could possibly be the Caliph, and Hasan bit his finger to see whether he was asleep. Then he thought that he must have been bewitched by some genie, some king of the genies, at least. But, little by little, as his orders were carried out and everyone behaved towards him with the very greatest respect, he became convinced that he was, indeed, Haroun Al Rashid.

He gave orders that the corrupt self-styled divines were to be thrashed; that all the pay of the soldiers was to be doubled; that everyone should be exempted from military service, that the River Tigris was to be dammed; that people who wanted to leave the city should be prevented and those who wanted to come in should be stopped ... In fact, such was the effect of his surrounding and the lack of any directing instinct upon him that, had his orders been carried out, the good order of the realm would have been seriously undermined.

The political advisers suggested that he should make alliances; the military commanders requested that he prepare for war; the merchants sent delegations pressing for higher prices; the citizenry petitioned for more liberal administration. The wise men counseled caution and this and that action. Hasan listened to all of them and was influenced now by one and now by another.

And all this happened within the space of a single day, between his waking up in the early morning following the soporific draught and the evening of the same day.

All the time the Caliph and Ja'far and Masrur watched their unwitting guest from a place of concealment which had been specially built for the purpose.

Finally, Haroun called his friends together and said:

"This is not the man whom we seek; one who will respond to

the opportunities and also the difficulties of power in such a way as to make the most of human life, discharging obligations and carrying on an enterprise for the good of all, including himself. Release him!"

So Hasan was again drugged, dressed in his old clothes, and taken to the cross-roads near his home, where he woke up sometime later shouting, "I am the Caliph, and demand that you obey me!"

When he was thoroughly awake, however, he was quite sure that it had all been a dream. From time to time after that, he used to think that that had been quite the most amazing sort of dream, that he had indeed lived in another reality. But he never was able to return to it.[44]

Now it is possible to perceive that the intent of the original story "Abu and the Caliph" was to emphasize the fact that the process of transformation of man's inner nature is much more sophisticated than just placing him in a somewhat "nobler" environment artificially constructed. A supposedly "nobler" environment may help to change one's manners. It does not, however, affect man's inner nature.

It is interesting to notice that Shakespeare also adapted the same tale in its uncorrupted version. He used the tale as the frame-story for his play "The Taming of the Shrew." In the frame-story, a mischievous nobleman tricks a drunken tinker named Christopher Sly into believing he is actually a nobleman himself. While drunk, Sly is carried to the lord's house and slipped into a fine bed. When he awakens, he is told that he is a great lord who has lost his memory, and his previous experiences were but a dream. Then a troupe of players arrives, and they are asked to entertain Sly with their performance. The players present the story of Katharina the Shrew.

---

[44] *Seeker After Truth*, Idries Shah (The Octagon Press, London, 1985, p. 49-51).

Sly falls asleep during the performance. At this point, he vanishes from the play without any explanation. Obviously, his future acts and his doing are simply ... irrelevant. The frame-story serves as a reference to the original tale of "Abu and the Caliph." In this way, Shakespeare points out that it is his play-proper that illustrates the process of the true transformation of man's nature. The "taming of the shrew" is a technical term describing a formula leading to the transformation of a raw human into a truly noble being. In the play, this is illustrated by the transformation that Katherina is going through.[45]

There are many versions of the story known as "A Magic Carpet." The version quoted below has been published by Idries Shah in the collection entitled *The Way of the Sufi*. It is this version that may be of help to modern physicists as they struggle with their search for the Theory of Everything.

There are two kingdoms in this story, the kingdom of Khorasan and the Unseen Land. The Unseen Land represents the invisible world. We may recognize that this tale illustrates the process leading to the entanglement of these two kingdoms.

> Once upon a time, there lived two brothers, Kasjan and Jankas. One day, the brothers decided to leave their home and seek their fortune somewhere else. They walked away from their home and it was not long before darkness separated them. By a strange coincidence, Kasjan, the younger brother, became the owner of a flying carpet.
> In the meantime, Jankas was snatched by a huge bird who dropped him on a street in a city in the kingdom of Khorasan.
> Kasjan commanded the carpet to take him to wherever his

---

[45] *Shakespeare for the Seeker*, *Vol. III*, W. Jamroz (Troubadour Publications, Montreal, 2013, p. 145).

brother was. At that very moment, however, Kasjan was thinking that his brother probably made himself a prince. The carpet heard this thought and delivered Kasjan to the palace of the king of Khorasan.

The king recognized in Kasjan the young man who was foretold to appear in the palace and who would be able to explain the mysterious behavior of the king's daughter who used to disappear every night and return in the morning and nobody knew how. Kasjan agreed to help and suggested that he should watch by the princess' bedside. In the middle of the night, a terrible spirit appeared, took the princess on his shoulder and they soared together through the ceiling. Kasjan ordered the carpet to take him where the princess had gone. In no time, Kasjan found himself in the Unseen Land. There was the princess walking with the spirit through forests of trees of precious stones and then through a garden of unknown plants of matchless beauty. At one point the pair stood by a lake whose reeds were shimmering swords. Kasjan overheard the spirit saying that these swords can kill spirits like him. At this very moment, Kasjan seized one of the swords and cut off the spirit's head. He took the princess back to the palace. Then he flew on his carpet to find Jankas, his still missing brother. When they returned to the palace, the princess was immediately smitten by the manly features of Jankas. The princess and Jankas were married. The king handed over the kingdom to them. Afterward, the king and Kasjan transferred themselves on the magic carpet to the Unseen Land, which became their joint kingdom.[46]

---

[46] This is an abbreviated version of the tale entitled "The Cap of Invisibility" from *The Way of the Sufi*, Idries Shah (The Octagon Press, London, 1989, p. 205).

At the beginning of the story, the link between the two kingdoms is corrupted. This is symbolically indicated by the presence of a terrible spirit. The link is there but it does not allow for constructive interaction between the two kingdoms. Quite to the contrary, the spirit interferes within the affairs of the kingdom of Khorasan. The instrument that is needed to establish a coherent link is presented as a "flying carpet." The story points out that the "flying carpet" functions as a faculty of the mind; it follows Kasjan's train of thoughts. At the beginning, Kasjan's thoughts are not precise, they are not refined enough. Only when his thoughts are focused on a clearly defined and constructive target, can the carpet discharge its function correctly and precisely. In other words, the carpet is that seemingly imaginary "operator" which allows Kasjan to bridge the two worlds.

Idries Shah's writings have also been used to help unravel the writings of other modern writers and poets. For example, the book entitled *Jorge Luis Borges: Sources and Illumination* contains an analysis of the works of Jorge Luis Borges, an Argentinian writer.[47] The book shows how Idries Shah's writings may help to get a better grasp of Borges' symbolic illustrations. For example, in one of Borges' short stories, there is a description of a mysterious "operator" that takes on the shape of the Semitic letter *Aleph*. In this story, Carlos Argentino, Borges' acquaintance, tells him about an Aleph that "is one of the points in space that contain all points." Carlos further explains that the Aleph is in one corner of the cellar in his house. Apparently, he has discovered it in his childhood. The Aleph, according to Carlos, is "the place where, without admixture or confusion, all the places of the world, seen from every angle, coexist." The author thinks that his acquaintance is simply a madman, but he decides to verify that strange story. Upon arrival at Carlos' house, he is told that he will have to lie on the tile floor in

---

[47] *Jorge Luis Borges: Sources and Illumination*, Giovanna de Garayalde (The Octagon Press, London, 1978).

the cellar and fix his eyes on the nineteenth step of the stairway. The author follows the instructions and sees the Aleph:

> ... Under the step, towards the right, I saw a small iridescent sphere of almost unbearable brightness. At first, I thought it was spinning; then I realized that the movement was an illusion produced by the dizzying spectacles inside it. The Aleph was probably two or three centimeters, but universal space was contained inside it, with no diminution in size. Each thing was infinite things, because I could clearly see it from every point in the cosmos. I saw the populous sea, saw dawn and dusk, saw the multitudes of the Americas, saw a silvery spider-web at the center of a black pyramid, saw a broken labyrinth (it was London), saw endless eyes, all very close, studying themselves in me as though in a mirror, saw all the mirrors on the planet (and none of them reflecting me), saw in a rear courtyard on Calle Soler the same tiles I'd seen twenty years before in the entryway of a house in Fray Bentos, saw clusters of grapes, snow, tobacco, veins of metal, water, vapor, saw convex equatorial deserts and their every grain of sand, saw a woman in Inverness whom I shall never forget, saw her violent hair, her haughty body, ... saw a copy of the first English translation of Pliny, saw every letter of every page at once, saw simultaneous night and day, saw a sunset in Querétaro that seems to reflect the color of a rose in Bengal, ... saw the Aleph from everywhere at once, saw the earth in the Aleph, and the Aleph once more in the earth and the earth in the Aleph, saw my face and my viscera, saw your face, and I felt dizzy, and I wept, because my eyes had seen that secret, hypothetical object

whose name has been usurped by men but which no man has ever truly looked upon: the inconceivable universe.[48]

It does not really matter much whether the above account is Borges' dream, or a record of his personal experience, or simply a poetical fantasy. The mechanism of connecting the visible world with the invisible follows the same mechanism as that described in "A Magic Carpet." Entering the Unseen Land with forests of trees of precious stones requires an operator to induce such a transformation in the human mind. In the experience described by Borges, the operator was in the shape of the Aleph. Borges' experience is a partially encrypted version of Ibn-Arabi's vision described in *The Bezels of Wisdom*. We may notice that Borges' *Aleph* is an equivalent of Ibn-Arabi's *Alif*; the *Alif* provided a hidden frame-story for Ibn-Arabi's vision. In Borges' description, instead of "words of wisdoms" there are very sophisticated images of those "wisdoms." The image of "universal space that was contained inside it, with no diminution in size" or the image of "a copy of the first English translation of Pliny[49] ... every letter of every page at once" – cannot be confined within any deterministically defined space, whether four-, eleven-, or higher-dimensional ones. These images belong to a qualitatively different realm that is beyond the reach of an ordinary rational mind.

It is through this sort of images that the readers are exposed to multiple views at the same time. Such exposure stimulates the brain to operate simultaneously in both the sequential and the holistic modes. The net effect of such an experience is the awakening of innate capacities of the human mind. When the innate capacities are activated, the conventional notion of time and locality breaks down.

---

[48] *The Aleph*, Jorge Luis Borges, translated by Andrew Hurley (Penguin Books, London, 1998, p. 129-131).
[49] *Natural History*, Pliny the Elder (an editorial model for encyclopedias comprising 37 books, published around 77 AD).

It is then that seemingly odd and improbable events, which are separated by large distances and taking place at different times, start to form a single and coherent narrative. In this way, the tales prepare the human mind for an encounter with higher states of consciousness, just like science fiction has prepared the human mind for an encounter with astrophysics and modern cosmology.

It looks like the illustrations of the invisible worlds have been changing over the centuries. Starting with symbolic words of wisdom and going to a magic country of precious stones growing on trees and concluding with the image of a placeless world. We may realize that these projections were reflections of changes within the human mind. Gradually, these projections have been converging towards the image that was being drawn in parallel by science.

Complex versus Real

> The complex numbers are the most beautiful things ever invented by man.
>
> (*Roger Penrose*)

So, how may "A Flying Carpet" and "Aleph" help physicists get out of the impasse in which they are presently locked? How are these tales relevant to the problem encountered in their search for the Theory of Everything?

The relevance of these tales to the current situation may be grasped by recalling the modus operandi of the human mind. Namely, the human mind may operate within two domains. One of the domains is accessible through the physical senses. This domain is within the reach of the rational mind. The other domain belongs to the Macrocosm. It is there that the overall matrix of the rational mind is located. The Macrocosm, however, is accessible only through supersensory faculties. This means that to fully comprehend the events registered by the physical senses, supersensory faculties are needed. However, these supersensory faculties are not active within the rational mind; they must be developed.

The supersensory faculties act as an operator allowing one to link together these two domains, the physical and the macrocosmic. Allegorically, such a situation may be represented by two kingdoms. One of these kingdoms represents the physical world. The other kingdom is placed within an unseen world, a sort of imaginary or magical world; it represents the world of the supersensory faculties. Tales that are set within such an allegorical scheme hold a hint about the *operator* that links together these two kingdoms. "A Flying Carpet" and "Aleph" are examples of such allegorical illustrations. The flying carpet and the Aleph – are symbolic representations of such an operator. And it is this observation that offers a hint to today's physicists.

Let us recall that the matrix of the matter is located within the lowest zone of the field of cosmic consciousness. This zone is just below the Planck length. Therefore, to be successful, any version of the Theory of Everything would have to include this particular zone of consciousness. Yet, the string theory is limited to the regions above the Planck length. In other words, it does not include the region belonging to that zone of consciousness. According to the hint provided in the tales, such an approach cannot lead to a satisfactory solution. It does not really matter how many extra dimensions are added to the string theory. As long as that specific zone of consciousness is not included, the overall approach is simply not adequate for a satisfactory description of "Everything." This may be compared to trying to fit a polygonal peg into a round hole. It does not matter whether it is a four-sided or eleven-sided peg – it will not fit completely into a round hole. As long as physicists insist on ignoring the field of consciousness as the overall operating framework, they will not be able to succeed in their quest to unravel "the mind of God."

In all previously tackled problems of physics, the elements of the solution belonged squarely to the physical world. Now, however, for the first time, the solution is placed within the invisible world; it is beyond the reach of the rational mind. This puts modern science in a rather awkward position.

As of today, physicists are divided on how to handle this situation. Some of them, which include those who call themselves "realists," have decided to go back to the 1920s and rewrite the foundation of quantum mechanics. A second group has embarked on a search for a link between quantum physics and a neurological model of consciousness. A third group has adopted a "stamp-collection" approach by devoting themselves to collecting more and more exciting and exotic astrophysical and cosmological data. A fourth group is focused on convincing the international community to invest tens of billions of dollars in constructing a new super-collider; they believe that with the help of a new super-collider they will be able to find the answer by exploring further the quantum world in the region located a bit closer to the Planck length. A fifth group still tries to reformat the string theory so that it could be verified experimentally.

The "breakthroughs" and "discoveries" that are being reported in various journals and magazines nearly every week – are fragmented pieces of these various efforts.

None of these approaches will provide a satisfactory solution.

Physicists will have to adjust their conceptual view of science and realize that their overall approach must be modified accordingly. As indicated by Zeno's paradox – a fuller understanding of the working of the universe is different from intellectual extensions of observed facts. It looks like the Theory of Everything needs to be set up within a different "frame-story" than the one which physicists have been dreaming about so far.

***

The critical factor needed to understand the quantum world is that consciousness is a form of energy. Consequently, and in agreement with Einstein's famous equation ($E = Mc^2$), the matter is also a form of consciousness. The relationship between matter and consciousness remains beyond the grasp of today's physicists.

According to the model of cosmic consciousness, the equivalence of consciousness (C) and matter (M) may be expressed by a simple formula[50]:

$$C = M (\alpha + i\beta)$$

In this formula, cosmic consciousness C is expressed in the form of a complex number. It is important to keep in mind that the symbol C used here applies only to the lowest zone of the field of consciousness, i.e., the zone that contains matter. The symbol M denotes the entire spectrum of observable matter.

The expression in the bracket, $(\alpha+i\beta)$, is a symbolic representation of the *operator* that links matter with consciousness. In other words, this "bracket" is an algebraic equivalent of the *flying carpet* or the *Aleph*. It is represented by a complex number. The real part (α) describes that part which manifests itself as observable matter. The imaginary part (β) describes that part of matter that remains in its oscillating invisible form. It is this part that has been identified by physicists as dark matter. It is this type of relationship that will have to be included in a satisfactory formulation of the Theory of Everything. In other words, in order to overcome the current impasse, physics will have to reach beyond the physical "nothingness."

We may notice that the above formula describing the relationship between consciousness and matter is an algebraic translation of the

---

[50] *A Journey through Cosmic Consciousness*, W. Jamroz (p. 141, see Note #1).

concept expressed by Omar Khayaam in the verse quoted earlier in this book (see Note #23). In that verse, Khayaam says that higher consciousness may be induced by emptying one's mind from shallow thoughts. Khayaam allegorically pointed out the relationship between matter and consciousness by referring to a centrifugal force ("Your head will turn and turn, vertiginously"). To access higher levels of consciousness ("Mounted the soaring Burak of their thoughts") one is required to sort out lighter substances from heavier ones. In other words, he refers to consciousness as ... a substance. Matter represents the heaviest or lowest substance along the gradient of consciousness. A symbolic illustration of such "emptying one's mind" was later on encrypted by Rumi in the choreography of the whirling dervishes.

The above formula indicates that consciousness and matter are coupled together. The ratio of that coupling is not fixed; it depends on the stage of the evolutionary growth of the universe. The ratio of the coupling, therefore, is quite a sophisticated parameter. This ratio cannot be measured; it can only be modeled and approximated. For example, at the beginning of the creation of the physical world, the imaginary part was dominant. At that time, there was no physical matter, and the entire space-time was filled-up with dark matter. At the present stage of the life of the universe, both components of the operator (α and β) are non-zero. Space-time is filled-up with both, dark matter and physical matter. There is a continuous flow of dark matter into the observable matter and, at the same time, the observable matter is being converted back into dark matter. However, the ratio of physical matter to dark matter is not constant. It changes as the universe is going through its life cycle.

In the previously discussed examples, complex numbers have greatly helped to analyze and solved various equations. However, complex numbers were not really needed. They became a useful tool for a variety of scientific and engineering applications. In these applications, they allow one to describe multiple relationships between various physical quantities. For example, complex numbers greatly simplify the analysis of voltage and current in electronic circuits. Voltage and current are real quantities; they both can be measured. In order to simplify the analysis, however, it is convenient to assume that one of them is an imaginary quantity. This allows to treat two of them together as a complex number. Then, it is possible to use a single equation to describe the multiple relationships that link them together. In addition to its usefulness, there is a certain elegance in unifying two real quantities into a single complex one.

In the case of consciousness and matter, however, the situation is quite different. Namely, complex numbers are unavoidable. They are an integral part of the relationship between these two quantities. They link together two different domains, the real and the imaginable, the visible and the invisible, the measurable and the unmeasurable. Specifically, the imaginary part of the complex function serves as a frame-story for the real part. Now we may realize that the entire sequel leading to the development of imaginary numbers was needed to prepare the rational mind for its encounter with the macrocosmic domain.

It may be said that complex numbers are an aspect of cosmic reality; they are part of nature. In other words, the Cosmos is "complex." Therefore, Roger Penrose's statement quoted at the beginning of this chapter is applicable only to the "real," i.e., the physical world. Indeed, in the physical world, imaginary numbers were *invented*. However, from the point of view of the "complex" Cosmos, the imaginary numbers were *discovered*. The complex numbers reflect the relationship between the physical world and the Macrocosm. This simple analogy allows to draw a line between scientists and mystics. Namely, scientists are dealing exclusively with the physical or "real" world. Mystics, on the other hand, are experts in the "complex" Cosmos.

The relationship between matter and consciousness is probably the most significant feature and the biggest secret of humanity. As we have seen, to perceive it – one needs to exercise one's imagination. Imagination provides a space in the mind into which inspiration may flow. But, as emphasized by Kingsley Dennis, this must be "true" and not "fantastical" imagination.

We can now realize that it was this relationship between cosmic consciousness and matter that was hidden within the Egyptian truncated pyramid and was waiting for the time when it could be recognized by the rational mind. It took many steps, which were separated by centuries before this "secret" could be discovered and comprehended. The invisible capstone of the Great Pyramid of Giza, the invisible letter in Ibn Al-Arabi's vision, the invisible book of Rumi's *Mathnawi*, the invisible frame-story of Shakespeare's plays – all of these are hints pointing towards this "invisible" relationship. In this relationship, the imaginary part acts as the main-frame for the visible.

All the episodes involving the Egyptian papyrus, Heron of Alexandria, Musa al-Khwarizmi, Omar Khayaam, Gerolamo Cardano, and Erwin Schrödinger – were milestones leading to the gradual discovery of this "secret" relationship. Finally, our Genie could be freed.

Therefore, I was not really surprised when, a few days after I wrote the above formula for the first time, a pyramidal crystal that was on my desk ... shattered; it turned into a small pile of yellow dust. This crystal was given to me by Santa. It was attached to the book that Santa gave me thirty years ago.

The Entangled Mind

> History is not an attempt to ascertain the truth, but a system of propaganda, devoted to the furtherance of modern projects, or the gratification of national vanity.
> (*John Bagot Glubb*)

The development of civilizations is a reflection of the changes which are taking place within the human mind. By directing the evolution of the human mind, the cosmic matrix is guiding the development of societies. As described in the previous chapters, the changes within the cosmic matrix are projected onto the physical world. These changes are manifested in the form of new concepts and ideas. In their symbolic forms, these concepts and ideas appear first in scriptures, poetry, and tales. At the next stage of their projection, the cosmic changes affect certain individuals, groups of people, and then entire nations. At the latter stage of the process, the macrocosmic adjustments are manifested as the appearances and disappearances of entire civilizations. However, it takes some time before these macrocosmic patterns are absorbed and digested by ordinary minds. Yet, it is these patterns that guide the evolution of civilizations. By looking at the development of civilizations, it is possible to get a glimpse of the working of the cosmic matrix.

Each civilization has a characteristic signature. There is something distinctly different about each of them. Or, to put it in another way, each civilization has contributed something to the formation of our present-day society. For example, Greece gave us drama, music, philosophy, and mathematics. Rome, on the other hand, introduced a code of law, an Empire-wide infrastructure, a network of roads and public works. We may recognize that these two empires developed features that were needed for the more effective functioning of future communities. This indicates that each empire had something to achieve, something to accomplish, and something to pass on to its successors. After discharging its function, it disappeared, and the next empire sprouted out somewhere else. In other words, the Phoenix which rose from its own ashes did not look like the previous one. It was somehow richer than its previous version. History does not repeat itself – it only seems like it does.

Such a view of civilizations is different from that offered by historians. Historians tend to interpret the disappearances or collapses of civilizations as the result of human shortcomings. This implies that there was a possibility to prevent their disappearances. This would be equivalent to blaming a man for ... dying. Instead, there is a natural transition from one stage onto the next stage of the process.

If we look at the world map, we can see that the various civilizations appeared in different geographical regions. They seemed to be somehow re-seeded from one place to another: Assyria, Egypt, Persia, China, Greece, Rome, the Moorish Empire, the Ottoman Empire, the British Empire, and so on. This observation leads to questions which, somehow, historians are not in a position either to pose or answer:

-How is the location of the seed of a new empire determined?
-Did each empire achieve its intended purpose?
-Is there a possibility within this process for an empire to not fulfill its intended function?
-What would be the consequences of an empire failing its intended role?

The answers to these questions can be found only by looking at the process from the macrocosmic perspective. As we have seen earlier in this book, they were provided in the third version of the

Arabian Nights. Namely, the seeds of new civilizations are planted according to an overall plan. According to this plan, every major region on the planet needs to contribute to humanity by serving as a … superpower. No empire in the past was able to discharge fully its evolutionary potential; however, some of the empires were more effective than others. The success or failure does not affect the empire's lifetime. However, an empire's failure makes it more challenging for its successor.

The interaction between the Macrocosm and the physical world is like that of a resonating circuit – just like that illustrated in Chladni's experiment. Depending on the frequency of oscillations and the location of the plate's attachment points, the grains of sand form different shapes. For each frequency and location of the attachment points, there is a specific geometrical pattern formed by the grains of sand. It may be said that the formed pattern is a characteristic "design" or "signature" of the particular combination of the parameters of the resonating plate.

The same mechanism applies to civilizations. Humanity is like grains of sand placed on a resonating plate. At different historical times, various nodes of the field of consciousness were purposely activated within select communities in designated geographical areas. In this manner, the various civilizations came into being.

The pattern projected from the Macrocosm contains the characteristic signature of a new civilization. For this pattern to be projected effectively, certain individuals must be present within that community. These individuals are part of the resonating circuit. They constitute the "receiving" group. In the language of quantum mechanics, it may be said that they are entangled to the cosmic matrix. They constitute the "entangled mind" of their community. Therefore, they are capable of perceiving and putting themselves in resonance with the projected design. As a result, their nature is changed. The mechanism of that change has been symbolically imprinted in the process of the metamorphosis of a caterpillar. The caterpillar's tissue contains cells that can resonate at frequencies which, when activated, destroy the restraining outward shape and convert the caterpillar into a butterfly; a new species appears.

We may think about the outward shape of the caterpillar as man's egotistic self-faculty which prevents the activation of subtle faculties. The self-faculty acts as a constraining shell. However, when the ordinary faculties of the mind are correctly realigned, then it is possible to overcome man's egotistic nature and form an advanced human. It is in this manner that the receiving group evolves into a "new" human species. An entangled mind is formed.

At the time of such transitions, there are intermissions or gaps in the macrocosmic projection. In other words, when the previous "design" gradually disappears, there is a delay and only then a new pattern appears. This may be seen by watching Chladni's experiment. When the attachment point is repositioned, first we see that the previous design disappears and there is no pattern at all. It takes some time for the grains of sand to overcome their inertia. After a brief period – a new pattern gradually appears. The gap serves as a cleansing operation; it allows for the erasure of the old pattern. Such cleansing is needed to correctly prepare the entire environment by removing elements that may interfere with the new design, just as the destruction of the caterpillar's outward shape is needed for the formation of the butterfly. In the case of civilizations, these gaps are the source of all sorts of turbulences and chaotic events that accompany such transitions. These disturbances are the manifestations of human raw nature when humanity is temporarily unrestricted by an operating cosmic pattern. They are like X-ray scans that show the true state of humanity. During those periods, the receiving groups serve as the pillars of their communities because they permanently absorbed the latest design. They act as a backup or a "safety valve" when the operating cosmic pattern is temporarily switched-off. In this way, they sustain their communities. The rest of the population, however, starts to behave in accordance with patterns which are at several removes backward from the current one. It may seem that a large part of the community temporarily regresses. These periods have been identified as the various historical "dark ages."

The receiving group and its location serve as the "attachment" points. It is through this group that a particular geographical area emulates the pattern which carries the "signature" of a new civilization. In this way, the seed of a new civilization is planted. However, at this stage of the process, the appearance of the new design is obscure to other members of that community.

At the next stage of the process, the projected pattern affects other members who are sensitive enough to detect the presence of a new "vibe" within their community; they are inspired by it. They form the outward shape of the new design. They are like the "grains of sand" that form the visible form of the pattern. However, to be "inspired" is not the same as being "changed." The inspired people may be completely unaware of the source of their ... inspiration. At this stage of the process, the projected pattern is partially distorted. The distortion is proportional to the depth of egotism of the "inspired" members of the community. This is the critical stage of the process as the degree of distortion determines the signature of the new civilization. The response of "inspired" members leads to certain undertakings, whether in science, arts, music, politics, business, or social activities. The "inspired" people become the decision-makers and the leading members of intellectual, artistic, and political elites; they are the celebrities of their community. Their activities contribute to the formulation of a new culture that becomes the signature of that civilization. It is through this group that the remaining part of the community becomes affected by the new design. At this stage of the process, the new signature is manifested as a new science, a new trend in arts, a new fashion, a new social norm, or a new political system. The development of quantum mechanics is one of the examples of that process. In this case, it was the group of "metaphysicists" from Copenhagen that performed such a function. However, this sort of influence is only the manifestation of external makeups. As illustrated in the tale of "Abu and the Caliph," such influences do not affect the very nature of people. They are simply a mark –not so much of evolving but rather– of civilizing people. Because, as described by Sir Geoffrey Howe, it is hardly a progress for a cannibal to use a fork and a knife.

There is always a limited time within which a given community is exposed to the newly projected design. When the assigned period of time expires, there is a change within the matrix that leads to the appearance of a new pattern. A new "receiving" group is formed in another geographical region. The seed of another civilization is planted. At the same time, the previous civilization gradually dies. And so on, till each major region on the planet has a chance to go through such an experience.

At this point, it is important to observe that the prime drive within the cycle of empires that we have seen so far has been domination and supremacy. This has been the driving force and this cycle's chief characteristic. During this cycle, communities have been "protected" by their empires. This is nicely summarized by Steven Weinberg in the following observation:

> "It is only a small exaggeration to say that, until the introduction of the post office, the chief service of nation-states was to protect their people from other nation-states."[51]

---

[51] *Dreams of a Final Theory*, Steven Weinberg (Vintage Books, New York, 1992).

Communities have been protected in the same manner as children are protected by parents. Therefore, the cycle of empires may be compared to the "childhood" of humanity. This has been an early phase of the process.

By drawing a parallel between the evolution of the human mind and the development of civilizations, it is possible to envisage the forces which will be forming the future world structure. In the process of the evolution of the human mind, the transition from "childhood" to "adulthood" requires the rearrangement and harmonization of the mind's natural faculties. In the case of the development of civilizations, this would be equivalent to the harmonization of multiple communities on a global scale. The first traces of this sort of harmonization may be discerned within the later stages of the current cycle of empires. The most successful empires were formed as an amalgam of people coming from many origins and building a new society. Moorish Spain was the first example of such an empire that appeared in the 8th century. In Moorish Spain, ethnic Spaniards, Jews, Arabs, and Persians developed and protected art, science, and culture during the Dark Ages of Europe. The appearance of the British Empire marked the next stage of such a harmonization. Britain provided a seedbed for the intermingling of the stocks and heritages of the Celts, Angles, Saxons, Romans, Normans, Vikings, and others. Just like Moorish Spain, Britain produced a singular society. The United States of America is the latest example of such a harmonization.

Therefore, it may be assumed that, at one point of the evolution of societies, the drive for domination will be replaced, at least partially, by a focus on global priorities. Global issues will become as important as local priorities. This means that the priority of the future world structure will be weighted between global and local interests. This would be a major qualitative change versus the current cycle of superpowers. It should be emphasized that the refocus on global issues will not be driven by a sort of altruistic motive. Quite to the contrary. The implementation of a global architecture is just a smarter and more effective way to secure the well-being of all involved. Barack Obama, a former president of the U.S., described this approach in the following way:

> Our motivation for erecting this architecture had hardly been self-less. Beyond helping to assure our security, it pried open markets to sell our goods, kept sea-lanes available for our ships, and maintained the steady flow of oil for our factories and cars. It ensured that our banks got repaid in dollars, our multinationals' factories weren't seized, our tourists could cash their traveler's checks, and our international calls would go through.[52]

In this context, it is interesting to observe that scientists have already been working in such a harmonized community for some time. It may be said that they formed the first example of such a "global" community. The modern scientific community was formed in Europe at the end of the 16th century. At that time, the members of this community realized that the laws of physics are universal. Regardless of race, religion, language – the laws of physic take on the same form and are applicable across the entire physical world. It was that realization that permitted the formation of a global community of like-thinking individuals.

It is important to notice that, with the appearance of the United States of America, the last remaining piece of land on the world map became the host of a superpower. It may be said that, with the foundation of the United States, each major territory on this planet

---

[52] *A Promised Land*, Barack Obama (Crown - Random House, New York, 2020, p. 328).

has experienced the taste of being an empire or a superpower. Or we may put it in another way: each territory had a chance to contribute to the development of our modern society. This may be a mark of the completion of the initial phase of the process. Namely, most human races had an opportunity to contribute to the formation of the modern world.

All these observations indicate that the current phase of the development of civilizations is coming to its end. It may be assumed that, presently, we are witnessing the first signs of the winding-down of the cycle of superpowers. The current state of the development of civilizations may be looked at as the beginning of the transition from a superpower-based cycle into a qualitatively different cycle. We may think about such a transition as going from childhood to adulthood. Just like a child is getting out of the protection of his parents, so the new world structure that is ahead of us – will not need the "supervision" of superpowers. Consequently, the new world order will be driven by a different force; a force that would be more closely aligned with humanity's ultimate purpose.

In the past, these sorts of transitions were marked by wars, political upheavals, plagues, floods, and other natural disasters. These were the various forms of "dark ages." The current situation bears all the signs that we are entering into such an unstable period. Suddenly and without any apparent reason, we are faced with a number of disturbing and stressful situations that are popping out all over the place at the same time. Without any obvious motive, large groups of people have started to behave as if they had been partially stripped off from their "civilizing coating" and show themselves in their raw and naked nature; their ways of thinking seem to have reverted to reasoning that was characteristic of the medieval times. It looks like some restricting and civilizing constraints have been temporarily lifted. It is the first time in our history that we are experiencing something like this on a global scale. This is an indication that a major transition is about to take place.

The biblical "flood" discussed earlier in this book may serve as quite a relevant reference point. In the context of a transition from one civilization to another, it is possible to unravel the allegorical meaning of the "flood," "ark," and "Noah's family." Namely, the "flood" marks a transition; the "ark" is a vehicle of transition, i.e., a specific pattern of the cosmic matrix; "Noah's family" represents a "receiving" group.

The main challenge of any transition is the fact that its modus operandi is beyond the grasp of ordinary intellect or logical reasoning. This type of transition is concerned with the working of heightened states of the human mind. These heightened states seem to follow the principles of the quantum world. Namely, the outcome of the transition is probabilistic but within specific boundaries. In other words, the development of civilizations is driven by cosmic imponderables. Furthermore, as the process advances, the degree of human responsibility becomes greater. The situation is very much the same as it is in the quantum world: the upper limit of the quantum world is determined by human inventiveness. A higher degree of responsibility is the "price tag" that comes with entering into "adulthood." "Adulthood" requires us to correctly discharge a specific function. It is at this point that human consciousness and free-will come into the equation.

There is no warranty that, at this time, humanity will meet its evolutionary challenge. This means that the outcome of the coming transition is not determined. It may be possible to envisage available options, but we cannot know if, as a society, we will be able to correctly fulfill our potential to warrant a transition to a more advanced cycle. In the case of failure, humanity will not be able to "mature"; it will have to stay longer within the cycle of superpowers.

The outcome of the coming transition will be determined by the quality of the present and future "inspired" groups. If their degree of readiness is adequate for the newly projected design, then it may be possible to absorb a qualitatively new evolutionary pattern. In such a situation, humanity will be able to enter onto a higher turn of the evolutionary spiral. To sustain that evolutionary state, in the coming years and decades more and more humans will have to develop perceptive skills which will allow them to act as constructive "influencers." Only then will it be possible to leave behind the "childish" cycle of superpowers.

The fact that such a transition has been initiated would indicate that there is a strong probability that the intended outcome is within human reach. Otherwise, we would not be able to envisage it.

Books by the same author

A Journey through Cosmic Consciousness, Troubadour Publications (2019)

A Journey with Omar Khayaam, Troubadour Publications (2018)

Shakespeare's Elephant in Darkest England, Troubadour Publications (2016)

Shakespeare's Sequel to Rumi's Teaching, Troubadour Publications (2015)

Shakespeare's Sonnets or How heavy do I journey on the way, Troubadour Publications (2014)

Shakespeare for the Seeker, Volume 4, Troubadour Publications (2013)

Shakespeare for the Seeker, Volume 3, Troubadour Publications (2013)

Shakespeare for the Seeker, Volume 2, Troubadour Publications (2013)

Shakespeare for the Seeker, Volume 1, Troubadour Publications (2012)

En español

Un viaje por la consciencia cósmica, Troubadour Publications (2020)

Un viaje con Omar Khayaam, Editorial Sufi (2020)

Shakespeare para el buscador (Completo: 4 volúmenes – versión Kindle), Editorial Sufi (2020)

El elefante de Shakespeare: en la Inglaterra más oscura, Troubadour Publications (2017)

Rumi y Shakespeare, Editorial Sufi (2016)

Shakespeare y su maestro, Editorial Sufi (2015)

Shakespeare para el buscador - Volumen 4, Editorial Sufi (2013)

Shakespeare para el buscador - Volumen 3, Editorial Sufi (2011)

Shakespeare para el buscador - Volumen 2, Editorial Sufi (2011)

Shakespeare para el buscador - Volumen 1, Editorial Sufi (2011)

En français

Voyage à travers la conscience cosmique, Troubadour Publications (2021)

Made in the USA
Columbia, SC
11 June 2021